TREES, SHRUBS, AND WOODY VINES OF GREAT SMOKY MOUNTAINS NATIONAL PARK

BY ARTHUR STUPKA

Park Naturalist, 1935-60

Biologist, 1960-64

THE UNIVERSITY OF TENNESSEE PRESS • KNOXVILLE • 1964

Dedicated to those members—past and present—of the University of Tennessee's Department of Botany whose names appear herein. Although very significant contributions have been made by other botanists, particularly R. H. Whittaker and W. H. Camp, the sum total of the writings, collections, and various miscellaneous efforts by members of the University of Tennessee's staff of botanists, extending over a long period of years, resulted in a wealth of information on that aspect of Great Smoky Mountains National Park which is its most distinctive attribute—the plant life.

Preface

This account deals with the trees, shrubs, and woody vines in Great Smoky Mountains National Park—native, naturalized, or escaped from cultivation.* Trees are defined as "woody plants having one erect perennial stem or trunk at least 3 inches in diameter at breast height (4½ feet), a more or less definitely formed crown of foliage, and a height of at least 12 feet" [Little, 1953]. Shrubs are woody perennials, smaller than trees, usually with several stems [Fernald, 1950]. Just as the differences between trees and shrubs are essentially matters of definition, so too the line of demarcation between woody and non-woody plants is not sharp. Therefore some entities, such as spotted wintergreen *(Chimaphila maculata)* and partridge-berry *(Mitchella repens),* while they are included by the present author, might be excluded if someone else was writing the text. Woody vines are woody plants whose stems require support, and which climb by tendrils or other means or which creep along the ground.

The present review is not intended to serve as a means of identifying the species of woody plants in the park, although occasionally data of that nature is given. The information presented here usually includes (1) the common and scientific names of the woody plants occurring in the area; (2) whether these are native or non-native; (3) whether common, rare, etc.; (4) the localities where the plant occurs, including habitat and altitudinal range;

---

* The woody plants growing in Twin Creeks, a residential area for government employees approximately 1 mile southeast of Gatlinburg, are excluded. A large variety of exotic species, planted there in the 1920's and 1930's by Louis E. Voorheis who then owned the land, continues to thrive; some of the species occur nowhere else in the park.

(5) data relating to the period of flowering; and (6) records of the biggest individuals of a number of the trees. Additional miscellaneous notes are included on occasions.

Information pertaining to Great Smoky Mountains National Park is obtainable from various sources [King and Stupka, 1950; Whittaker, 1952, 1956; Stupka, 1960, 1963], particularly the current issue of the official booklet on the area [U. S. Department of the Interior, 1959]. Regular text type distinguishes native species in the text; smaller type is used for non-native species, for species occurring in the immediate vicinity but not within the park's boundaries, for those that no longer occur in the area, and for some whose status is questionable.

For their generosity in furnishing the photographs that appear in this book I wish to thank Mr. R. M. Schiele, Director of the Schiele Museum of Natural History, Gastonia, North Carolina; Mr. James E. Thompson, Knoxville, Tennessee; Dr. William F. Hutson, Chicago, Illinois; Mr. Walter M. Cline, Chattanooga, Tennessee; the Division of State Information, Nashville, Tennessee; and the National Park Service (Great Smoky Mountains National Park).

Contents

List of Illustrations

Introduction

SOURCES OF INFORMATION

Under "References Cited" are all the written (published and unpublished) sources of information I examined in preparing the present account. In addition, all the specimens of woody plants collected within or in the immediate vicinity of Great Smoky Mountains National Park that are now in the University of Tennessee's herbarium or in the herbarium in the Great Smoky Mountains National Park were inspected—altogether totaling more than 1,000 sheets.

With the exception of a number of introduced plant species, practically all those discussed in the present review are represented by specimens in the herbarium in Great Smoky Mountains National Park, the herbarium in the University of Tennessee, or in both places. Fortunately a number of the genera including some of the more "difficult" ones had been examined previously, and determinations had been made by recognized authorities.* In the section entitled "Previous Botanical Studies" reference is made to some of the most important sources of information, and of these the following recur most frequently as origins of basic data:

H. M. Jennison. "A Classified List of the Trees of Great Smoky Mountains National Park" [1938] and "A Preliminary

---

* W. P. Adams, *Ascyrum;* L. H. Bailey, *Rubus;* C. R. Ball, *Salix;* B. V. Barnes, *Populus;* W. H. Camp, *Gaylussacia, Leiophyllum, Rhododendron* (in part), and *Vaccinium;* S. Hu, *Philadelphus;* G. N. Jones, *Amelanchier* and *Tilia;* K. Lems, *Leucothoë, Lyonia,* and *Pieris;* W. H. Lewis, *Rosa;* W. E. Manning, *Carya;* W. L. McAtee, *Viburnum;* E. J. Palmer, *Crataegus;* H. L. Sherman, *Prunus;* H. T. Skinner, *Rhododendron* (in part); B. M. Speese, *Smilax;* B. L. Wagenknecht, *Fothergilla;* and F. W. Woods, *Ilex.*

1

Catalog of the Flowering Plants and Ferns of the Great Smoky Mountains National Park" [1939a]. Except for some minor changes, all of the 1938 list formed a part of the one prepared in 1939. Neither of these briefly annotated lists was published.

A. J. Sharp. "A Preliminary Checklist of the Trees in the Great Smoky Mountains National Park" [1942a] and "A Preliminary List of the Woody and Semi-Woody Shrubs and Vines Occurring in Great Smoky Mountains National Park" [1942b]. Both these lists were revised by Fred H. Arnold in 1945, largely to conform to *Standardized Plant Names* [Kelsey and Dayton, 1942]; these unpublished lists have brief annotations. In 1956 Sharp et al. prepared "A Preliminary Checklist of Monocots in Tennessee"; in 1960 this was followed by "A Preliminary Checklist of Dicots in Tennessee"—both issued in mimeographed form by the University of Tennessee. In these lists an asterisk is used to designate those entities believed to occur in Great Smoky Mountains National Park.

R. E. Shanks. "Reference Lists of Native Plants of the Great Smoky Mountains" [1954]. Shanks' mimeographed compilation is a listing of the higher plants occurring in six "broad physiognomic types" of forests. Instead of the "numerous specific vegetation types which might be recognized," Shanks simplifies the complex vegetation pattern by reducing the categories to (1) cove hardwood forests, (2) hemlock forests, (3) northern hardwood forests, (4) spruce-fir forests, (5) closed oak forests, and (6) open oak and pine stands. In 1961 Shanks favored the author with a 19-page typewritten list (unpublished) entitled "Woody Plants of the Great Smoky Mountains" in which 255 species plus a number of subspecies are enumerated. His unpublished lists of the park species of *Vaccinium* [1947], *Crataegus* [1953], and *Rubus* [1957] proved very helpful in the preparation of the present report.

R. H. Whittaker. "Vegetation of the Great Smoky Mountains" [1956]. One of the sections in Whittaker's monographic study combines species whose "distributional centers are relatively close together in relation to environmental gradients. Species thus grouped tend to occur together in many stands. . . ." Some of these natural aggregations, which he calls "classes" or "unions," occupy dry ("xeric") situations while others are to be found in intermediate ("mesic")* situations. This information is usually given in the present account under the appropri-

---

* The prefix "sub," meaning "somewhat," is also used. Thus "submesic" denotes "somewhat mesic."

ate species. Reference is also made to some of Whittaker's information on altitudinal ranges as indicated in his population charts [ibid.]. Whittaker kindly favored me with his opinions of the blueberry *(Vaccinium* spp.) and huckleberry *(Gaylussacia* spp.) populations in the park as a result of correspondence he carried on with the late W. H. Camp. What appears to be an instance of natural hybridization between *Rhododendron maximum* and *R. catawbiense* was also reported to me by Whittaker.

Arthur Stupka. Nature Journal, Great Smoky Mountains National Park. This is an unpublished recording of phenological data, occurrence of plants and animals in various places and at various altitudes, some climatological observations, etc. It covers the period from mid-October 1935 through the year 1962. An appreciable amount of the information relating to time of flowering and to altitudinal range, as given in the present report, was derived from this journal.

Roland M. Harper. "Preliminary List of Southern Appalachian Endemics" [1947]. This is the source of information on that subject.

With few exceptions, the standard used for nomenclature pertaining to the common and scientific names of trees and arborescent shrubs is E. L. Little's *Check List of Native and Naturalized Trees of the United States (Including Alaska)* [1953]. For shrubs and woody vines the authority ordinarily used is the 8th edition of *Gray's Manual of Botany* [Fernald, 1950], exceptions being the determinations and publications of E. J. Palmer relating to *Crataegus,* the determinations and publications of L. H. Bailey relating to *Rubus,* and the determinations and publications of W. H. Camp relating to some of the entities comprising *Vaccinium.* There are a few additional digressions. The phylogenetic arrangement of families and genera is according to Fernald [ibid.], but the species are presented alphabetically except in *Rubus* (segregated into raspberries, dewberries, and blackberries), *Rhododendron* (segregated into deciduous and evergreen species), and *Rhus* (segregated into sumacs and poison ivy).

PREVIOUS BOTANICAL STUDIES

The Southern Appalachian region has attracted botanists since "about the time of the Revolution" [Harper, 1947], but that part

of it which we now call Great Smoky Mountains National Park
—located between the Little Tennessee River on the west and
the Big Pigeon River on the east—had largely been bypassed by
persons interested in botany until after 1925. This was due, for
the most part, to its inaccessibility.

Dr. A. Gattinger, an emigrant from Germany who arrived in
East Tennessee in 1849 and spent 15 years in that part of the
state, appears to have been one of the first to collect plants in that
area. From a perusal of his writings [1887, 1901] and from his
biography [Oakes, 1932] it cannot be determined as to how many,
if any, of his journeys were made into areas now included within
Great Smoky Mountains National Park. Unfortunately his collection of some 4,000 species of plants that had been acquired
by the University of Tennessee in 1890 was destroyed in the
Morrill Hall fire in 1934; I have not examined the Gattinger specimens in the A. W. Chapman collection that were acquired in part
by Vanderbilt University and in part by the Missouri Botanical
Gardens.

One of Gattinger's correspondents was Dr. Albert Ruth, superintendent of schools in Knoxville, Tennessee. Ruth is known
to have collected rather extensively in what is now the park and
vicinity, and some of his material is now in the United States
National Herbarium, the herbarium of the New York Botanical
Garden, and elsewhere. Part of Ruth's herbarium was purchased
by the University of Tennessee shortly after the 1934 fire; only a
few species of trees and shrubs are represented, these bearing collecting dates in 1888 and 1890.

Another of Gattinger's correspondents was T. H. Kearney who
served as assistant botanist for the University of Tennessee Experi-
ment Station in 1892-93. Galyon [1928b] writes that during the
years 1890-93 Kearney "made extensive collections of the trees
and shrubs of Eastern Tennessee. . . . As far as we know, there is
no published paper showing the results of his works." The collections he contributed to the University of Tennessee Herbarium
were lost in the 1934 fire. That was also the fate of Samuel M.
Bain's specimens—Bain had been a former botanist at that insti-
tution [ibid.].

H. C. Beardslee, Jr., son of Dr. Henry C. Beardslee, botanist
of the State of Ohio, along with Oberlin College students C. A.
Kofoid and Clarence E. Hemingway collected plants in the Great

Smoky Mountains during the summer of 1891. Some, or all, of these collections became a part of the herbarium in Oberlin College.*

In their report on "The Southern Appalachian Forests," Ayres and Ashe [1905] describe forest conditions over much of the area now included within Great Smoky Mountains National Park, their "basins" or "districts" being comparable to watersheds or major portions of watersheds. Twenty-one such regions include park or sections of park lands, and for each of these information is given as to acreage (in square miles), acreage cleared, acreage burned, acreage in forest, topography, soils, agricultural value, timber trees (approximately 25-30 species are involved), etc. The percentages of the various timber trees comprising the total stand is given for about half the regions. In his letter of transmittal dated March 7, 1904, Henry Gannett describes this as "a report upon the examination of the forest conditions of a large area of the Southern Appalachian Mountains, made in 1900 and 1901, by Messrs. H. B. Ayres and W. W. Ashe . . . at the joint expense and under the joint supervision of the Geological Survey of North Carolina, represented by Prof. J. A. Holmes, State geologist, and the Bureau of Forestry of the Department of Agriculture, represented by Mr. Gifford Pinchot, of this office. . . ."

Another joint study entitled "Forest Conditions in Western North Carolina," made by J. S. Holmes [1911], forester to the North Carolina Geological and Economic Survey, assisted by W. B. Willey and A. W. Williamson, forest assistants in the United States Forest Service, "takes up a new phase of this subject, and treats of the present condition of the forest and of forest industries, with their economic relation to the people of the region and to the State as a whole. It is primarily for the owners of forest land to furnish them with information as to the proper management of their forest holdings." The forest and economic conditions of Swain, Haywood, and 14 other counties in western North Carolina are described as observed in 1909.

L. R. Hesler, emeritus dean of the University of Tennessee's College of Liberal Arts and formerly head of the Department of Botany, has pursued his studies on the fungi of the Great Smoky Mountains and vicinity for over 40 years. His publication of

* Information supplied by Dr. George T. Jones of Oberlin College, Oberlin, Ohio, in his letter to me dated January 18, 1963.

books and numerous scientific papers in the field of mycology represents the results of a dedicated interest; his list of park fungi [Hesler, 1962] includes 1,975 species. Throughout his long career with the University of Tennessee he was instrumental in attracting a number of outstanding botanists to the staff of the Department of Botany—particularly H. M. Jennison, A. J. Sharp, S. A. Cain, and R. E. Shanks—whose combined activities, especially in the realm of published accounts of the local flora, served to mark an era. It proved to be a most fortunate circumstance that in the early stages of the development of a National Park whose grandeur and uniqueness is based largely on its botanical features that there would be a source of information so conveniently located and made up of such distinguished personalities.

Harry Milliken Jennison was associated with the University of Tennessee from 1922 until his death in 1940. He was given a two-years' leave of absence from his teaching duties in the Department of Botany, beginning in 1935, during which time he served in the Great Smoky Mountains National Park as a wildlife technician under the Civilian Conservation Corps; his chief duties were the collection, identification, and preparation of plant specimens. Whenever possible the collections were made in triplicate, with one for the park's herbarium, one for the park's exchange collection, and one for the University of Tennessee's herbarium. In this endeavor he was usually assisted by an enrollee from one of the CCC camps in the park. By March 1937 approximately 4,000 sheets (specimens of plants) representing about 1,000 species had been deposited in each of the three categories.* In November 1935 his monthly report [Jennison, 1935b] included a list of the "Trees of the Great Smoky Mountains Park."

During the summers of 1938 and 1939 Jennison served in the park in the capacity of ranger-naturalist, and when time permitted he continued to augment the collections in the herbarium. In 1938 he prepared "A Classified List of the Trees of Great Smoky Mountains National Park" which, with minor alterations, formed part of the more comprehensive "A Preliminary Catalog of the

---

* Those who assisted with the work up to that time included A. J. Sharp (480 specimens); Lane Barksdale (435); D. C. Bain (200); R. J. Fleetwood (200); S. L. Wallace (200); John R. Raper (175); T. G. Harbison (110); and H. K. Svenson (50). Jennison, assisted by Glen Smith or some other enrollee, accounted for 2,270 specimens.

Flowering Plants and Ferns of the Great Smoky Mountains National Park" which he wrote in 1939.* His illustrated general account entitled "Flora of the Great Smokies" [Jennison, 1939b] was published in the *Journal of the Tennessee Academy of Science* in 1939.

In his description of *Rubus jennisonii*, a new blackberry from Great Smoky Mountains National Park, the eminent botanist and horticulturist Liberty Hyde Bailey [1945] wrote: "Dedicated to the memory of Dr. Harry M. Jennison (1885-1940) of the University of Tennessee, devoted teacher, ardent collector, choice companion." An obituary [Cain and Hesler, 1940] containing additional information appeared in the *Journal of the Tennessee Academy of Science*.

Both "Flowers of the Great Smokies," by Jesse M. Shaver, and "The Trees of the Great Smokies," by R. S. Maddox, appeared in the April 1926 issue of the *Journal of the Tennessee Academy of Science*. These are brief general accounts that relate to an area of appreciably greater size than the approximately 500,000 acres covered by the National Park.

In December 1927 Willa Love Galyon read a paper entitled "The Smoky Mountains and the Plant Naturalist" before a meeting of the American Nature-Study Society, in Nashville, Tennessee. This was published in the April 1928 issue of the *Journal of the Tennessee Academy of Science* [Galyon, 1928a]. Galyon's thesis entitled "Check List of the Trees and Shrubs of Eastern Tennessee" [Galyon, 1928b] was submitted in May 1928 to the graduate committee of the University of Tennessee.† At that time no development had yet taken place in the newly-authorized National Park, and its approximate boundaries were unknown except by a few individuals. Therefore the omission of reference to this area is understandable. The brief annotations occasionally make reference to "mountains" or "tops of mountains," but otherwise no information is given as to altitudinal range. Data on occurrence

---

* These briefly annotated unpublished reports were later used by A. J. Sharp in preparing his "A Preliminary Checklist of the Trees in the Great Smoky Mountains National Park" and "A Preliminary List of the Woody and Semi-Woody Shrubs and Vines Occurring in Great Smoky Mountains National Park," both of which were written in 1942.

† In partial fulfillment of the requirements for the degree of Master of Arts. It was not published.

and habitat is mostly generalized. However, with allowances made for the changes in taxonomy that have taken place in the past 35 years, most of the park's trees and shrubs are included. Cain's initial report on the plant ecology of the park is based largely on the list prepared by Galyon [Cain, 1936].

In 1927 Stanley A. Cain, while on the staff of Butler University, began botanizing in Great Smoky Mountains National Park. In 1930 two of his reports appeared in Volume 1 of *Butler University Botanical Studies:* "Certain Floristic Affinities of the Trees and Shrubs of the Great Smoky Mountains and Vicinity" (pp. 129-150), and "An Ecological Study of the Heath Balds of the Great Smoky Mountains" (pp. 177-208). In 1931 there appeared in the *Botanical Gazette* (Vol. 91) his "Ecological Studies of the Vegetation of the Great Smoky Mountains of North Carolina and Tennessee. I. Soil reaction and plant distribution" (pp. 22-41). In 1935 Cain joined the faculty of the University of Tennessee where he remained until 1946; during this interval his writings included the following:

1935. "Ecological Studies of the Vegetation of the Great Smoky Mountains, II. The quadrat method applied to sampling spruce and fir forest types." *American Midland Naturalist,* Vol. 16, pp. 566-584.

1936. "Ecological Work on the Great Smoky Mountains Region." *Castanea (Journal of the Southern Appalachian Botanical Club),* Vol. 1, pp. 25-32.

1937. Cain et al. "A Preliminary Guide to the Greenbrier-Brushy Mountain Nature Trail, the Great Smoky Mountains National Park." The University of Tennessee, Knoxville. Mimeographed, 43 pp., including cover page, map, and appendices.

1938. Cain and A. J. Sharp. "Bryophytic Unions of Certain Forest Types of the Great Smoky Mountains," *American Midland Naturalist,* Vol. 20, pp. 249-301.

1943. "The Tertiary Character of the Cove Hardwood Forests of the Great Smoky Mountains National Park." *Bulletin of the Torrey Botanical Club,* Vol. 70, pp. 213-235.

1944. *Foundations of Plant Geography.* Harper, New York, 556 pp.

1945. "A Biological Spectrum of the Flora of the Great

Smoky Mountains National Park," *Butler University Botanical Studies,* Vol. 7, pp. 11-24.

This considerable contribution served greatly to enrich the literature on the plant ecology of the area; its aggregate represents the enduring record of a gifted scientist.

Aaron J. Sharp, on the staff of the Department of Botany at the University of Tennessee since 1929, has been one of the most active botanists in our area. He served as head of the department for a period of 10 years, beginning in 1951. The papers he has published, some in collaboration with other scientists, number over 100 (1963); more than half are devoted to bryophytes and other non-flowering plants, the remainder cover a wide range of subjects with emphasis on the flora of Mexico and reviews of botanical literature. In the summer of 1934 he substituted for Dr. H. M. Jennison by accepting the assignment of wildlife technician for the recently-organized Civilian Conservation Corps in Great Smoky Mountains National Park. Sharp also served as a seasonal ranger-naturalist in this park during parts of the summers of 1940, 1941, and 1942. On numerous occasions he brought his classes and visiting botanists to the park, and during the annual Spring Wildflower Pilgrimage he has played a very active part. After the destruction of the University of Tennessee's herbarium in the 1934 Morrill Hall fire, Sharp played a leading role in the collecting of plants in many parts of Tennessee so that now (1964) the University's floral collections are among the best in the South. In 1956 Dr. Hu named *Philadelphus sharpianus,* a plant that Sharp had collected in the park, in his honor [Hu, 1956]—one of several such distinctions Sharp has acquired.

It was during the summer of 1942 that Sharp prepared two checklists based partly on his personal knowledge and collections of the local flora and partly on checklists previously compiled by the late Dr. H. M. Jennison—one, "A Preliminary Checklist of the Trees in the Great Smoky Mountains National Park," and the other, "A Preliminary List of the Woody and Semi-Woody Shrubs and Vines Occurring in the Great Smoky Mountains National Park." These lists are briefly annotated. In 1945 Fred H. Arnold, who was then the regional forester in the Richmond, Virginia, office of the National Park Service, revised both lists in accordance with the directive whereby Sudworth's *Check List of the*

Forest Trees of the United States (1927) was to be followed as the authority for the common names of the trees, while *Standardized Plant Names* (2nd edition, 1942) was to be the authority for the common names of shrubs (and herbs) and the scientific names of all plants.* These lists made no reference to time of flowering and were not published. In the present report these lists are referred to as "Sharp, 1942a" and "Sharp, 1942b."

With the late R. E. Shanks, Sharp collaborated in writing a "Summer Key to the Trees of Eastern Tennessee" [1947]. In 1956 he collaborated with R. E. Shanks, J. K. Underwood, and Eleanor McGilliard in the preparation of a mimeographed "A Preliminary Checklist of Monocots in Tennessee," a 33-page report "presented in order to permit collectors . . . to make corrections and suggestions regarding the species included and their ranges." In 1960 this was followed by "A Preliminary Checklist of Dicots in Tennessee," by Sharp, R. E. Shanks, H. L. Sherman, and D. H. Norris—a mimeographed report of 114 pages "prepared as part of the study on the vascular plants of Tennessee supported by Grants Nos. G-1478 and G-4446 from the National Science Foundation." In both these latter reports an asterisk is used with all species and subspecies known to occur in Great Smoky Mountains National Park; these unpublished checklists are referred to in the present report as "Sharp et al., 1956" and "Sharp et al., 1960." In 1962 Sharp collaborated with C. C. Campbell, W. F. Hutson, and H. L. Macon in the preparation of an illustrated (color) book entitled *Great Smoky Mountains Wildflowers* and in an enlargement of the book in 1964 [Campbell et al., 1962, 1964].

I wish to acknowledge Dr. Sharp's cooperation in the preparation of various parts of this present report and to thank him for his generous assistance with plant determinations, the search for pertinent reference material, and for his advice and comments.

---

* By a National Park Service directive dated December 29, 1959, this was modified as follows: Common and scientific names of trees were to be in accordance with E. L. Little's *Check List of Native and Naturalized Trees* [1953], while common and scientific names of all other flowering plants were to be according to *Standardized Plant Names*, 2nd edition, 1942. In September 1963 when the National Park Service questioned the use of *Standardized Plant Names* as the authority to follow in botanical nomenclature I revised the names of most shrubs and herbaceous plants in accordance with *Gray's Manual of Botany* [Fernald, 1950].

His review of this manuscript is but one of the many favors for which I am very grateful.

Dr. W. H. Camp, who was then at Ohio State University, began visiting the Great Smoky Mountains area in the late 1920's. In 1931 his article entitled "The Grass Balds of the Great Smoky Mountains of Tennessee and North Carolina" appeared in the *Ohio Journal of Science*.* By 1936 when he published an account of a visit to the park entitled "On Appalachian Trails," his deep interest in some of the members of the heath family (especially *Vaccinium, Gaylussacia, Leiophyllum,* and some of the *Rhododendrons)* was well advanced. A copy of his five-page letter to Forester Frank H. Miller on the subject of the blueberries and huckleberries of this area, dated November 22, 1937, is in the park files. In 1938 Camp's "Studies in the Ericales III. The genus *Leiophyllum"* was published in the *Bulletin of the Torrey Botanical Club.* In 1942 his papers "On the Structure of Populations in the Genus *Vaccinium"* and "A Survey of the American Species of *Vaccinium,* Subgenus Euvaccinium" appeared in *Brittonia.* In 1945 *Brittonia* was again the medium in which Camp published "The North American Blueberries with Notes on Other Groups of *Vacciniaceae."* His studies on the American beech appeared in 1951. Camp made fairly extensive collections of ericaceous and other plants in the park, and in his papers he made frequent reference to this area.

H. S. Pepoon, a physician from Chicago, Illinois, spent some weeks botanizing in the park and vicinity in the spring of 1931. In a letter dated March 16, 1935, addressed to Park Superintendent Eakin, Pepoon reported "some 1500 species" resulting from his 1931 visit. It is not known, however, what areas of the park and vicinity Pepoon investigated as his notes are often rather vague. Since his reported occurrence of such species as black spruce *(Picea mariana)* from "Le Conte above 5000 feet," Carolina hemlock *(Tsuga caroliniana)* from "upper rocky mountains near

---

* An article on the same subject entitled "A Forest Enigma," by Paul M. Fink, appeared in *American Forests* in the same year (1931). Beginning in 1936 B. W. Wells published five articles on the subject of the origin of Southern Appalachian grass balds [1936a, 1936b, 1937, 1946, 1956]. W. H. Gates' theory pertaining to the origin of high-altitude treeless areas appeared in 1941. One of the most comprehensive reports on this subject was published by Mark in 1958.

summits," water oak *(Quercus nigra)* from "lower Roaring Fork,"
Ohio buckeye *(Aesculus glabra)* from "woods, occasional," and
various other entities has not been verified, and since a number
of statements pertaining to the status of park plants appear to
be based upon insufficient investigation, I have not used this list
as a source of information.*

One of the basic undertakings realized during the years when
the Civilian Conservation Corps was active in Great Smoky Mountains National Park was the preparation of a vegetation type map.
This was begun in May 1935, with Senior Foreman (Forester)
Frank H. Miller in charge. Field work by type-mapping crews
spanned a period of more than three years (1935-38); Miller remained on the job through 1941. The finished map is available
for examination in the Sugarlands Visitor Center. In September
1938 a report by Miller to the Superintendent included "Brief
Narrative Descriptions of the Vegetative Types in the Great Smoky
Mountains National Park"; this 17-page account is based on a
modified version of the Society of American Foresters' *Forest
Cover Types of the Eastern United States* [1932].

The writer arrived in the area on October 14, 1935, to become the first Park Naturalist in Great Smoky Mountains National
Park, a position I was to hold for 25 years. Much of the information on the flowering periods of the park plants, along with data
on altitudinal ranges, comes from the nature journal I maintained.† Writings which resulted from this work include the *Great
Smoky Mountains National Park Natural History Handbook*
[Stupka, 1960] and *Notes on the Birds of Great Smoky Mountains National Park* [Stupka, 1963].

Royal E. Shanks served on the staff of the Department of
Botany, University of Tennessee, in 1940-41 and again from 1946
until his death in 1962. His wide interests, especially in the fields
of ecology, climatology, and taxonomy, are reflected in a number
of published articles and in various mimeographed lists and keys
pertaining to the Great Smoky Mountains flora. In 1947 he collaborated with A. J. Sharp in the preparation of a "Summer Key

* This list, on 3 x 5 cards, was donated to Great Smoky Mountains
National Park in February 1947 by Mrs. Dorothy Lexau of Gatlinburg.
It is filed with botanical data in the Sugarlands Visitor Center.

† One copy in the Sugarlands Visitor Center (in the park) and one in
the Interior Department Library, Washington, D. C.

to the Trees of Eastern Tennessee" [Shanks and Sharp, 1947]. In 1952 there appeared his "Checklist of the Woody Plants of Tennessee." His publications in *Ecology* include "Plotless Sampling Trials in Appalachian Forest Types" [Shanks, 1954b], "Climates of the Great Smoky Mountains" [Shanks, 1954c], and "Altitudinal and Microclimatic Relationships of Soil Temperature under Natural Vegetation" [Shanks, 1956].

With Frank W. Woods, Shanks made an extensive survey of the kinds of trees that are replacing the blight-killed chestnuts in Great Smoky Mountains National Park. This became the subject of a brief report in the *Journal of Forestry* [Woods and Shanks, 1957] and of a comprehensive account in *Ecology* [Woods and Shanks, 1959].

In April 1954 the Botany Department of the University of Tennessee made available "Reference Lists of Native Plants of the Great Smoky Mountains," a 17-page mimeographed compilation by Royal E. Shanks [1954a]. In 1956 and again in 1960 Shanks was one of the authors who contributed to the mimeographed preliminary checklists of Tennessee monocots (33 pages) and dicots (114 pages), respectively—both issued by the Botany Department. Another mimeographed item which Shanks prepared was a "Provisional Key to Tennessee Species of *Rubus*," in 1957 (6 pages). All these proved to be very valuable sources of information in the preparation of the present publication.*

In September 1947 Shanks presented me with a typewritten "Key to the Species of *Vaccinium* in the Great Smoky Mountains (adapted from Camp, 1945)." Again in September 1953 he favored me with a penciled list entitled "Great Smoky Mountains National Park *Crataegus*" on which appears those species of hawthorns, recently identified by E. J. Palmer, that are represented in the University of Tennessee Herbarium. Finally in 1961 he presented me with a typewritten copy of a 19-page list entitled "Woody Plants of the Great Smoky Mountains" which he prepared and which includes 255 species and a number of varieties. His untimely death in 1962 removed an exceptionally gifted botanist whose friendly cooperation endeared him to all who knew him [Hunt, 1962].

S. Glidden Baldwin of Danville, Illinois, served in the park as

* Further reference is made to some of these in the section entitled "Sources of Information."

a seasonal naturalist during part of the summer of 1946. His interest in locating the largest individuals of the various kinds of trees in the area resulted in a list of such specimens which he submitted to *American Forests Magazine* and which appeared in that periodical's subsequent information on the subject [Dixon, 1961]. This data is given where it applies in the present publication. Dr. Baldwin published two articles on the subject: "Photographing Big Trees in the Smokies" [1948a] and "Big Trees of the Great Smokies" [1948b].

The doctoral thesis of Robert H. Whittaker, University of Illinois (1948), is entitled "A Vegetation Analysis of the Great Smoky Mountains." This title was changed to "Vegetation of the Great Smoky Mountains" when it was published in 1956. This report represents an outstanding contribution by a talented plant ecologist. All who have more than a casual interest in the botanical features of the park will benefit by reading this scholarly presentation; for the serious student it constitutes a very important reference.

Harlan P. Kelsey, member of a committee of five appointed by Interior Department Secretary Hubert Work to investigate areas in the Southern Appalachian region that appeared to have National Park caliber, published a brief article entitled, "Unique Flora of the Great Smoky Mountains National Park" [1949]. Previously [1942], Kelsey and William A. Dayton comprised the editorial committee that prepared *Standardized Plant Names,* "a revised and enlarged listing of approved scientific and common names of plants. . . ."

In her book *Deciduous Forests of Eastern North America,* E. Lucy Braun [1950] has a brief description of the forests of Great Smoky Mountains National Park.

H. J. Oosting and W. D. Billings collaborated in an article entitled, "A Comparison of Virgin Spruce-fir Forest in the Northern and Southern Appalachian System" [1951]. Billings had previously [1937] collaborated with S. A. Cain and W. B. Drew in the preparation of a "Winter Key to the Trees of Eastern Tennessee." In 1957 Billings collaborated with A. F. Mark in a report on "Factors Involved in the Persistence of Montane Treeless Balds."

Fred C. Galle, horticulturist on the staff of Callaway Gardens, Pine Mountain, Georgia, began his studies of the azaleas on

Gregory Bald in the early 1950's. In January 1963 he favored the writer with a report on that vari-colored assemblage of these shrubs.

"The Beech Gaps of the Great Smoky Mountains" is the subject of a study made by Norman H. Russell [1953].

Vernon C. Gilbert, Jr., of the National Park Service was awarded his Master of Science degree by the University of Tennessee on the basis of his thesis, "Vegetation of the Grassy Balds of the Great Smoky Mountains National Park" [1954].

In his article entitled "In Search of Native Azaleas," Henry T. Skinner [1955] discusses the azaleas he found on Gregory Bald and elsewhere.

Replacement of chestnut in the Great Smoky Mountains of Tennessee and North Carolina is the subject of studies made by Frank W. Woods and Royal E. Shanks [1957, 1959].

"Ground Vegetation Patterns of the Spruce-fir Area of the Great Smoky Mountains National Park" is the subject of a study made by Dorothy L. Crandall [1958].

During the summers of 1958 and 1960, Robert S. Lambert of Clemson College, South Carolina, was engaged in a study of the logging operations that had taken place on lands now included in Great Smoky Mountains National Park. His 1958 activities were by contract with the National Park Service; his research in 1960 was financed by the Great Smoky Mountains Natural History Association. The reports are filed in the Sugarlands Visitor Center. In 1961 Lambert published an article entitled "Logging on Little River, 1890-1940."

The Great Smoky Mountains National Park was one of 10 areas studied by George S. Ramseur (University of the South, Sewanee, Tennessee) in his survey of "The Vascular Flora of High Mountain Communities of the Southern Appalachians" [1960].

In 1960 Duke University awarded W. B. Schofield the degree of Doctor of Philosophy based, in part, on his study entitled "The Ecotone between Spruce-fir and Deciduous Forests in the Great Smoky Mountains." This thesis has not been published (1963).

Harold L. Hoffman of Gatlinburg, Tennessee, prepared a "Check List of the Vascular Plants of the Great Smoky Mountains National Park" [1962]. It is "based on the collections on file in the Great Smoky Mountains National Park herbarium and the Tennessee State herbarium at the University of Tennessee,

Knoxville, Tennessee." The 44-page mimeographed list is without annotations.

LOGGING OPERATIONS IN PRE-PARK DAYS

Jennison [1939b], in his article on the "Flora of the Great Smokies," stated that "considerable timber has been cut, but within the confines of the Park this practice is now a thing of the past and, besides, the fact remains that within the Park's boundary there is upwards of 200,000 acres of virgin hardwood forest." That acreage, representing approximately 40 percent of the Great Smoky Mountains National Park, was arrived at by conferences with Frank H. Miller [1938] who was then engaged in the preparation of a vegetation type map of the area.

Lambert [1958], who compiled a list of 18 companies engaged in logging operations on lands now within the National Park, gives the following account of the two periods through which the local logging operations passed:

> The first or early period ended about the beginning of the twentieth century. It was characterized by small enterprises managed by local people and financed by local capital, which engaged in selective cutting of the most valuable timber of the day, particularly poplar, cherry and ash. Although these operators sometimes went deep into the forest in search of a particularly fine specimen, they generally stayed close to the creeks and other more accessible places. They also peeled some tanbark. Typical local methods of the day were used to get the logs out, including the use of animal and water power. The saw mills were usually of the small portable variety, and the product of the cutting was taken to the nearest railroad by wagon.
>
> The later period saw the coming of large-scale operations to the Park area. The growing demand for lumber and paper and the development of improved methods of obtaining and manufacturing wood brought outside capital and management into the area. To the managers of these operations, the Smokies represented just one section of the Southern Appalachians, and were simply an extension of ventures begun in West Virginia and Kentucky. With the advent of big-scale lumbering came the logging railroad which carried wood and lumber out from, and supplies in to areas long considered inaccessible. With the use of the latest logging equipment it was possible to cut all the merchantable timber within the boundaries and to sell it not merely for lumber and bark, but for use in the newer pulp and chemical industries. With these methods the only timber of any consequence that was left was that which grew in thin stands on rocky slopes from which

no profit could be expected. The mills were larger and were located in fixed positions on the edges of the timber.

The Little River Lumber Company whose mill and offices were located at Townsend, Tennessee, approximately 3 miles below the forks of Little River, was the largest operator in the area; the year 1939, when that firm ceased its timber-cutting activities in the park, marked the end of an era since it brought to a close all such practices on lands which hereafter would be dedicated to the preservation of all natural objects. "The Little River Lumber Company estimated that some 560,000,000 board feet of logs were manufactured by the mill at Townsend. In addition, unmeasurable quantities of timber were consumed in construction work at the mill, on the railroad, and in the camps, while some spruce was shipped directly in the log. The timber cut by the small farmers and selective loggers before 1900 is of unknown quantity, but it seems reasonable to speculate that the basin yielded as much as 1,000,000,000 feet of timber before cutting stopped altogether" [Lambert, 1961]. Lambert [1958] estimates that the activities of the various other companies operating within the confines of the "park" would approximately double that figure. This would "not include timber shipped in the log, that which was cut only for tanbark in the early days, that cut in the woods but never brought to the mill, that destroyed in operations by machinery and fire, or that used for construction and fuel."

Extensive logging was carried on in the Big Creek watershed, in Cataloochee Creek, along the Oconaluftee River and its main tributaries, in those major watersheds whose waters now empty into the Fontana Reservoir, and elsewhere. Much of this was done during the first 25 years of the present century. Fires were of rather frequent occurrence; these were sometimes caused by sparks from logging engines or could be attributed, directly or indirectly, to the timber-cutting operations. The prevalent practice, by farmers, of burning off the land was also a factor.

Information supplied to Lambert [ibid.] by the late D. H. Tipton, the last president and for many years general superintendent of the Little River Lumber Company, reveals the kind and quantity of trees that were cut at that mill during the period of its operations, 1903-39; the figures represent thousands of board feet:

Poplar*	60,076
Chestnut	46,687
Basswood	27,763
Maple	21,868
Oak	21,620
Buckeye	19,353
White Pine	14,592
Birch	10,778
Yellow Pine	9,962
Ash	6,136
Peawood (Silverbell)	5,935
Cherry	5,326
Miscellaneous	14,070
TOTAL FOR THE ABOVE	264,166

"These figures do not reveal . . . the role which hemlock played in the firm's total operations. Hemlock accounted for slightly more than half of the logs cut at the mill" [ibid.]. Although most of the hemlock was sold for pulp, the lumber that was manufactured (about 132,000,000 feet) represented "a quantity approximately equal to the production of poplar, chestnut and basswood" [ibid.].

---

* Refers to yellow-poplar *(Liriodendron tulipifera)* and, in all probability, includes the cucumbertree *(Magnolia acuminata)*.

TREES, SHRUBS, AND WOODY VINES
OF GREAT SMOKY MOUNTAINS NATIONAL PARK

A SYSTEMATIC ACCOUNT

FIR—*Abies* Mill.

FRASER FIR—*Abies fraseri* (Pursh) Poir.

This coniferous needle-bearing tree is confined to the Southern Appalachian region. Between 5000 and 6000 ft. it is usually associated with red spruce *(Picea rubens),* in general there being a greater percentage of fir with increase in elevation. Above the 6000 ft. altitude it forms almost pure stands. Its range in the park is from near Cosby Knob and Mt. Sterling Ridge west along the higher uplands to a point about 3 miles west of Clingmans Dome on the main divide. On north-facing slopes and in sheltered ravines some firs may occur down to approximately 4000 ft. (West Prong, Little Pigeon River), but usually the lower limit of this species lies between 4500 and 5000 ft. One of the best displays of this forest is along the ½-mile trail to the summit of Clingmans Dome.

The small, inconspicuous, cone-like "flowers" appear in late spring. The cones mature in the autumn, but red squirrels may begin feeding on the seeds as early as July.

The largest Fraser fir in the park, located approximately 100 yards west of the summit of West Peak of Mt. Le Conte, is 7 ft. 11 inches in circumference, 44 ft. in height, and has a spread of 25 ft. (1961).* Ordinarily mature trees are under 40 ft. in height, but occasionally much taller specimens are reported.

"Windthrow damage is characteristic of much of the high altitude fir forest, which is shallowly rooted. Following the overthrow of the canopy trees, fir seedlings develop rapidly, producing pole stands of even-aged trees. This cyclic type of reproduction

---

* A record specimen, according to the American Forestry Association [Dixon, 1961].

21

appears characteristic of the high altitude stands" [Crandall, 1958].

The discovery of the balsam wooly aphid, *Chermes picea,* on Mt. Sterling in August 1963 represents a very serious threat to the Fraser fir in Great Smoky Mountains National Park. This insect, known to have been introduced from Europe in 1908, has brought about the destruction of great numbers of fir trees in Canada, the northeastern United States, and in the Pacific Northwest. Control measures, undertaken in the Great Smoky Mountains in November 1963, are continuing.

The Fraser fir, also known as "balsam," is readily distinguished from the red spruce with which it often occurs. The fir is the only evergreen tree with upright cones. Unlike the cones of spruces, hemlocks, and pines, fir cones disintegrate in the autumn leaving a slender spike-like core on the tree. The fir often has a scattering of blisters on the bark, and its aromatic needles are blunt. (The cones of the red spruce are pendulous, blisters are absent on the bark, and the sharp-pointed needles lack the strong resinous odor.)

HEMLOCK—*Tsuga* (Endl.) Carr.

EASTERN HEMLOCK—*Tsuga canadensis* (L.) Carr.

This is the only native hemlock in the park.* Above 5000 ft. its occurrence is uncommon and rather spotty, but below that altitude this large tree is a common component of the forest preferring the ridges at elevations of 3500-5000 ft. and the shaded habitats along streamcourses, in cool ravines, and on north slopes in the lower areas. Shanks [1954a] grouped it with the trees that are dominant in hemlock, cove hardwood, and northern hardwood forests. Whittaker [1956] placed it in the mesic tree class "centered in cove forests below 4500 ft." Near the turn of the century, Ayres and Ashe [1905] estimated that hemlock comprised 60

---

* The Carolina hemlock *(Tsuga caroliniana)* has been planted in Gatlinburg, Tenn., but there is no record of this species from within the park. It occurs in the northeastern part of Tennessee (Carter County) [Shanks, 1952], in a number of places in North Carolina (including Highlands and in the Linville Gorge), and in Virginia, South Carolina (Caesar's Head and Table Mountain), and Georgia (Tallulah Falls) [Coker and Totten, 1934].

percent of the forest that covered the bottoms in the Cataloochee district.

The time of flowering is April (lower altitudes) and May (higher altitudes).

About 1 mile west of Brushy Mountain, along Surry Fork at an altitude of 3450 ft., grows an eastern hemlock measuring 19 ft. 10 inches in circumference—the largest specimen of this species on record [Dixon, 1961].* Throughout the park are many hemlocks measuring 15 ft. in circumference, or over. In the vicinity of the big yellow-poplar (23 ft. in circumference) along the Indian Camp Creek Trail to Maddron Bald is a hemlock measuring 17 ft. 3 inches in circumference.

Gilbert [1954] reports a 15 ft. circumference hemlock at 5700 ft. altitude on Mt. Sterling Bald. Whittaker's [1956] population chart likewise gives 5700 ft. as the uppermost limit for this species. Hemlocks have been noted up to 5500 ft. on Inadu Knob, Thunderhead, Spruce Mountain, and near Charlies Bunion and to 5400 ft. along the Bull Head Trail to Mt. Le Conte. As already stated, this tree is rather scarce above 5000 ft. in the park.

Before the National Park became a reality, hemlock was cut in extraordinary quantities by lumber companies operating in the area. Lambert [1958] writes that approximately 33,000,000 ft. was harvested at the Smokemont, North Carolina, mill by the Champion Fiber Company from 1920 to 1925. On the Tennessee side, the Little River watershed "yielded as much as 1,000,000,000 ft. of timber before cutting stopped altogether—hemlock accounted for over fifty percent by volume of the timber cut, although much of it was turned out as lath rather than lumber" [Lambert, 1961].

One of our most attractive trees, the eastern hemlock, lacks the stiffness so characteristic of some of the other cone-bearing species. Its cones, averaging ¾ inch in length, are appreciably smaller than the cones of the native fir and spruce or of any of the native pines.

* Latest measurement, made October 1958. A photograph of this tree appears on page 14 of the *Great Smoky Mountains Natural History Handbook* [Stupka, 1960].

SPRUCE—*Picea* Dietr.

1. NORWAY SPRUCE—*Picea abies* (L.) Karst.

This European species is highly restricted in its occurrence in the area. Shortly after the 1925 conflagration that destroyed much of the forest cover in the Richland Mountain and Charlies Bunion regions, the Champion Fiber Company obtained and planted thousands of young Norway spruces near the headwaters of Kephart Prong, mostly at altitudes of 4000-5000 ft. A few trees have also been noted along Thomas Ridge, above Cherokee Orchard, and in the Cherokee Indian Reservation on Soco Bald close to the park.

2. RED SPRUCE—*Picea rubens* Sarg.

At the 4000-6000 ft. elevations over much of the eastern half of the park, especially where lumbering operations failed to pene- trate, this tall-growing conifer occurs abundantly. "The best- developed and most extensive virgin forests of spruce-fir in the southern Appalachians are in the Great Smoky Mountains National Park" [Oosting and Billings, 1951]. Good examples are to be seen along the road from Newfound Gap to Clingmans Dome, along all the trails to Mt. Le Conte, and elsewhere. Along the main Great Smokies range, both the red spruce and Fraser fir have their westernmost extension approximately 3 miles west of Clingmans Dome.*

Although the bulk of the red spruce forest lies between 4500 and 6000 ft., there are mature specimens as low as 3500 ft. near the headwaters of the West Prong of the Little Pigeon River and possibly elsewhere. A careful search in the forests at still lower altitudes reveals small, widely-scattered individuals whose pres- ence can be explained only by wind-borne seeds.

In the western half of the park the elevation of the Great Smokies range is appreciably less than in the eastern half. Never-

* The statement that the range of the red spruce extends south in the mountains to Georgia [Illick, 1928; Coker and Totten, 1934; Harlow, 1942] appears to be in error. In a letter dated February 10, 1942, Forest Super- visor H. B. Bosworth of the Pisgah-Croatan National Forests informed me that "the most southern occurrence of these species [red spruce and Fraser fir] is on the south side of Tennessee Bald adjacent to the Blue Ridge Park- way at a point approximately eight miles northwest of Brevard, North Carolina." This station is slightly further south than the outliers of these conifers that grow about 3 miles southwest of Clingmans Dome.

theless most of the mountains along the 20-mile ridge from Silers Bald southwest to Parson Bald are high enough for the growth of red spruce, yet no spruce is to be found there nor did this species occur there in historic times. An explanation proposed by Whittaker is based upon the assumption that a warm ("xerothermic") period prevailed for a time following the last glaciation [Flint, 1947].

Climatic warming during the xerothermic period sufficient to displace the lower limits of spruce-fir forests upward from 4500 ft. to approximately 5700 ft. would account for present distribution of these forests. . . . the spruce-fir forests extended farther south during glaciation than at present—how much farther can scarcely be guessed. During the last xerothermic period they were pushed upward to 5600-5800 ft. elevation and were pushed off the tops of the lower peaks south of Clingmans Dome. As the climate cooled again, the forests advanced down the slopes from the higher northeastern peaks where they had found sanctuary and reoccupied the land above 4500 ft. . . . The spruce forests should have been moving southwest along the ridge from Clingmans Dome in the 4,000 years since the peak of the xerothermic period (Flint, 1947), but are perhaps retarded or halted by the extensive beech forests of Double Spring Gap [Whittaker, 1956].

The red spruce is usually in "flower" during May or June. Cones mature at the end of the first season but may persist well into the following year. Red squirrels may begin feeding on the seeds by late July. Among birds, red crossbills and red-breasted nuthatches feed most persistently upon these seeds.

"According to Verne Rhoades, formerly Executive Secretary, North Carolina Park Commission, a detailed cruise of 20,000 acres of unburned virgin spruce in the Great Smoky Mountains showed an average of 35 trees to the acre having a diameter of 8 inches or more, containing an average of 25 to 30 cords to the acre when 6-inch trees were included. Red spruce in these mountains is known to have reached an age of 300 years, a diameter of 57 inches at breast height, and a height of 162 feet, many trees being found that exceeded 100 feet in height" [Korstian, 1937]. Along the trail to the pumping station that serves the facilities at the Clingmans Dome Parking Area is the stump of a red spruce that measures 3 ft. 6 inches in diameter; this stump reveals 353 annual rings.

According to the American Forestry Association [Dixon, 1961], the largest red spruce on record, measuring 14 ft. 1 inch in circumference (breast high), grows in Great Smoky Mountains National Park.*

In his article on the red spruce in the Southern Appalachian Mountains, Korstian [1937] writes that "out of an original area of virgin spruce forest estimated at 1 million acres, less than one-tenth remains, of which about 50,000 acres is under Federal and State ownership and a considerably smaller area is privately owned." Lambert [1961] reported that the "outbreak of the First World War created a demand for spruce for use in airplane construction. The estimated spruce sawlog stumpage held by the [Little River Lumber] company was 15,000,000 board feet; of this over 9,000,000 feet were cut and shipped in 1917 and 1918." This cutting was confined to the higher altitudes between Mt. Collins and Double Springs Gap. From 1920 to 1925 the Champion Fiber Company's mill at Smokemont, North Carolina, cut over 43,000,000 ft. of red spruce [Lambert, 1958]. Above Baxter Creek, in the Big Creek watershed, an operator cut 1,500,000 ft. of spruce for the war effort, while at the western limits of this tree in the park—at the headwaters of Forney Creek—cutting was at the rate of 40,000 ft. per day [ibid.]. Had the movement for the creation of a National Park in these mountains been delayed, the destruction of this splendid boreal forest would have been inevitable.

PINE—*Pinus* L.

1. SHORTLEAF PINE—*Pinus echinata* Mill.

Sharp [1942a] regarded this tree as being "not uncommon," while Miller [1938] said that "the stands are few and scattered on the North Carolina side but on the Tennessee side of the park the greater portion of the western end is under this type." It is partly due to the difficulty encountered in one's efforts to distinguish this from the similar-appearing pitch pine *(P. rigida)* that its status

* Verne Rhoades, in a letter dated July 2, 1958, informed the superintendent, Great Smoky Mountains National Park, that this tree which he saw and reported "stands in Bear Pen Gap at the head of the Right Fork of Three Forks." Later, in the course of a search that failed to relocate the big tree, James G. Hollandsworth and party discovered a red spruce on Break Neck Ridge with a circumference of 13 ft. 9 inches; height, 106 ft.; and spread, 45 ft. This, then, appears to be the present "champion."

remains somewhat unclear. This pine inhabits the lower altitudes; Sharp [1942a] stated that it may venture to 4000 ft. There are specimens from Rich Mountain, Cades Cove, Elkmont, and Greenbrier.

As in the other native pines, "flowering" takes place in April or May.

The cones and the bark of this and the pitch pine *(P. rigida)* are quite similar in appearance. The leaves (needles) of the shortleaf pine tend to be more slender and flexible than those of the pitch pine and their color a dark blue-green; the leaves of pitch pine are usually a more yellowish-green in color. While the shortleaf pine has its leaves in bundles of two or three, the pitch pine's leaves are in bundles of three.

2. LONGLEAF PINE—*Pinus palustris* Mill.

At approximately 2550 ft. altitude on Noland Creek, a few miles northwest of Bryson City, North Carolina, grow six longleaf pines, the largest of which measures 5½ inches in diameter and approximately 35 ft. in height. The location is 4 or 5 miles above Fontana Reservoir and only a few feet above the truck trail. Willard Jenkins of Bryson City who planted these trees in about the year 1937 told me that they were obtained approximately 40 miles north of Pensacola, Florida, by Mr. Rust who was then the proprietor of a considerable tract of land on Noland Creek—later to become a part of Great Smoky Mountains National Park.

In June 1963 I talked with Mr. Jenkins and obtained the above information; it was then that I last examined these trees and noted that a number of faster-growing yellow-poplars *(Liriodendron tulipifera)* were encroaching upon and overtopping the exotic longleaf pines. The latter trees, with needles averaging a foot in length, bore no cones.

3. TABLE-MOUNTAIN PINE—*Pinus pungens* Lamb.

This tree, one of the endemic species of the Southern Appalachians, is fairly common on dry south-facing slopes especially at middle altitudes. Shanks [1954a] includes it among the characteristic trees of the open oak and pine stands, while Whittaker [1956] places it in the xeric tree class. Specimens have been recorded as low as 1400 ft. elevation (Abrams Falls vicinity), but ordinarily the best stands are at altitudes of approximately 3000-4500 ft. On Andrews Bald, at 5800 ft., a stunted Table-Mountain pine marks the highest range of this species. It also occurs, although in limited numbers, on Inadu Knob, on Rocky Spur of Mt. Le Conte, and near Double Springs Gap—all at 5500 ft. or over. This pine grows on or near the main state-line divide on Spence Field Bald, Gregory Bald, and Parson Bald. Excellent stands of it are to be found growing near the summit of Greenbrier Pinnacle, along the Bull Head Trail to Mt. Le Conte, and above the upper Sugarlands Valley.

What appears to be the largest recorded specimen of this pine grows at an altitude of 4600 ft. beside the trail to the summit of Greenbrier Pinnacle, less than 300 ft. below the fire tower. In January 1961 this tree measured 6 ft. 11 inches in circumference, stood approximately 45 ft. high, and had a spread of about 40 ft.*

The Table-Mountain pine is readily identified by its massive cones armed with stout curved spines; these usually occur in whorls of three or more cones which persist for many years. The needles, occurring in pairs, are stout, stiff, twisted, and sharp-pointed.

4. PITCH PINE—*Pinus rigida* Mill.

This is a common tree in open oak and pine stands [Shanks, 1954a], especially at lower altitudes. Whittaker [1956] places it in his xeric tree class, along with *Pinus virginiana, P. pungens, Quercus coccinea,* and *Q. marilandica.* It ranges from elevations below 1000 ft. (Abrams Creek) to near 5700 ft. on Inadu Knob near the state-line ridge. Some specimens occur on Gregory Bald, Parson Bald, and on Greenbrier Pinnacle.

The largest pitch pine on record is one that grows in the Elkmont Campground at approximately 2200 ft. elevation; it measures 9 ft. 3 inches in circumference at breast height, is about 75 ft. high, and has a spread of approximately 45 ft.† Along the Cable Mill race in Cades Cove is another fine specimen measuring almost 8 ft. in circumference.

This tree is distinguished from the shortleaf pine *(P. echinata)* with some difficulty. Both trees may have three needles in a

---

* The largest specimen reported by Dixon [1961] is 6 ft. 10 inches in circumference (Chattahoochee National Forest, Georgia).

† The record given by Dixon [1961] is for a specimen measuring 8 ft. 3 inches in circumference (in New Jersey).

bundle (ordinarily, however, the needles of shortleaf pine occur in pairs), and the bark and cones are quite similar. (See shortleaf pine.)

5. EASTERN WHITE PINE—*Pinus strobus* L.

Whereas this large tree is rather scarce in some areas, it is common in others. According to Ayres and Ashe [1905], the eastern white pine comprised 20 percent of the forests growing in the bottomlands of the Cataloochee district. Shanks [1954a] groups this tree with species comprising the open oak and pine stands, while Whittaker [1956] includes it in his subxeric tree class "centered in oak-chestnut heath and drier oak-chestnut forests."

The eastern white pine grows from the lowest altitudes (Abrams Creek) to approximately 5000 ft. Specimens occur on Spence Field Bald, Gregory Bald, and Parson Bald. Some fine trees are to be seen along the short Pine-Oak Nature Trail in Cades Cove, including one that is 10 ft. 6 inches in circumference. Along Noland Creek at approximately 2000 ft. and along one of the lower tributaries of that stream are plantings of white pines made some years prior to the acquisition of that land for Great Smoky Mountains National Park. Another planting of these trees was made in the vicinity of Junglebrook, between Gatlinburg and Cherokee Orchard. In one of his reports Miller [1938] states "It has been recorded that logging operations for this species started as early as 1875 on the Carolina side of the park."

This is the only native pine with five needles to the bundle. The spindle-shaped cones hang in clusters from the ends of the higher branches.

6. VIRGINIA PINE—*Pinus virginiana* Mill.

On poor soils in lowland areas, particularly in abandoned fields, this is a common tree. Shanks [1954a] groups it with trees of the open oak and pine stands, while Whittaker [1956] includes it in his xeric tree class. Ordinarily it is quite scarce above 3000 ft., but Jennison [1939a] reports it at 5500 ft. along the Appalachian Trail near Double Springs Gap which represents its uppermost limit in the park.

The needles of the Virginia pine are grouped in pairs. From the Table-Mountain pine *(P. pungens),* which also has paired needles, the Virginia pine can be differentiated by its smaller cones with slender prickles. The needles of the shortleaf pine *(P. echinata)*, which may have two or three needles in a bundle, usually average an inch or more longer than those of the Virginia pine.

ARBORVITAE—*Thuja* L.

1. NORTHERN WHITE-CEDAR—*Thuja occidentalis* L.

Jennison [1938] reported that the northern white-cedar was "introduced about homesites and still persisting in a few localities" in the park. The natural distribution of this tree remains outside the park; its southern limit, according to Sharp, is believed to be at Falls Creek Falls State Park, Tennessee.*

2. ORIENTAL ARBORVITAE—*Thuja orientalis* L.

A few specimens of this cultivated ornamental "were planted near dwellings in our area," according to Jennison [1938]. This and the preceding species are of rare occurrence in the park and, in all probability, will eventually disappear.

JUNIPER—*Juniperus* L.

EASTERN REDCEDAR—*Juniperus virginiana* L.

This small to medium-size tree occurs rather infrequently in the park. Where limestone prevails, such as in Cades Cove and Whiteoak Sink, it is fairly common; just outside the park in Tuckaleechee and Wears Coves it is plentiful. The eastern redcedar is quite scarce on the North Carolina side of the park and over wide areas there it is absent. Ordinarily it is to be found below 2000 ft. altitude, but a specimen has been noted at 4500 ft. on the Bote Mountain road, and Gilbert [1954] records a small tree at 4730 ft. on Parson Bald.

If mild temperatures prevail in January, this tree may shed pollen as early as the latter part of that month (earliest: January 26, 1937), but usually it is in February or March before that event

---

* According to Coker and Totten [1934], this tree "comes no farther south than North Carolina, where it occurs only on limestone soil along Cripple Creek and Linville River, on the headwaters of the New River in Alleghany and Ashe counties, and near Jefferson and Sparta."

may be expected. Myrtle warblers are fond of the berry-like fruits produced by the redcedar and may be found wintering where these trees grow in appreciable numbers.

FAMILY ✦ *GRAMINEAE*

CANE—*Arundinaria* Michx.

SWITCH CANE—*Arundinaria tecta* (Walt.) Muhl.

This shrub is the only member of the grass family *(Gramineae)* to be regarded as one of our native woody plants. Its distribution in the park is quite localized, occurring as dense clumps along some streamsides and in moist situations below 2000 ft. Over wide areas it is scarce or absent. Places it has been noted include Abrams Creek, Little River near Elkmont, and Dudley Creek near Gatlinburg. It is our only native bamboo.

Specimens of an Asiatic relative, *Pseudosasa japonica,* were reported growing along Grassy Branch in the park, close to Gatlinburg, by H. L. Hoffman in June 1960. At that time these exotic plants were said to be 6 to 8 ft. tall.

FAMILY ✦ *LILIACEAE*

BEARGRASS—*Yucca* L.

ADAM'S-NEEDLE*—*Yucca smalliana* Fern.

Since this striking plant was occasionally transported from one place to another, its status is rather difficult to determine. Specimens are to be found in the vicinity of some former homesites, but then there are plants whose place of growth leads one to believe they established themselves without aid from mankind. It is rather uncommon in the park where it grows in sandy well-drained soils at low altitudes. One, growing in an old field near Smoke-

* Sharp et al. [1956] list this as *Y. flaccida*. Jennison [1939a] also includes *Y. filamentosa*.

mont, at approximately 2200 ft. elevation, appears to be at the upper limit of its range. This plant grows in Cades Cove, along lower Abrams Creek, above Fontana Reservoir, near Gatlinburg, and elsewhere.

During some years the Adam's-needle begins to bloom at the end of May. The usual peak of flowering is June 10 to 20.

GREENBRIER—*Smilax* L.

1. SAWBRIER—*Smilax glauca* Walt.

This woody vine is common at low and middle altitudes "in sandy or rocky soils in open woodlands, thickets and wet open places" [Jennison, 1939a]. Shanks [1954a] lists it as one of the typical tall "shrub" species of the open oak and pine stands, while Whittaker [1956] includes it as a member of his subxeric heath union.* The sawbrier has been noted at 3000 ft. altitude (trail to Greenbrier Pinnacle), and it probably occurs at somewhat higher elevations in the park.

The flowering season is normally from late May (earliest: May 27, 1934) to middle June. From the other two woody species of *Smilax* in the park, this one is readily separated by the whitish (glaucous) undersurface of its leaves.

2. BRISTLY GREENBRIER—*Smilax tamnoides* var. *hispida* (Muhl.) Fern.

This potent brier is fairly common, ranging from the lower altitudes to 5500 ft. (vicinity of Silers Bald). The flowering season is May and June. The bristly greenbrier has numerous black needle-like prickles which distinguishes it from the other two woody species of *Smilax* in the park.

3. COMMON GREENBRIER—*Smilax rotundifolia* L.

In woods and thickets and along roadsides this common brier is to be found from low altitudes to approximately 5000 ft. (Spence Field Bald). Shanks [1954a] lists it among the plants characteristic of cove hardwoods, closed oak forests, and open oak and

---

* Along with *Kalmia latifolia, Lyonia ligustrina,* and *Vaccinium constablaei.*

pine stands, while Whittaker [1956] includes it in his submesic shrub union "centered in oak-hickory and oak-chestnut forests."

This, like the other greenbriers, blooms in May and June. The tough stem is armed with formidable prickles.

FAMILY ✦ *SALICACEAE*

WILLOW—*Salix* L.

1. WHITE WILLOW—*Salix alba* L.

This foreign species is rare in the park. One growing near an old homesite in the Fighting Creek valley was designated as the white willow by C. R. Ball.

2. WEEPING WILLOW—*Salix babylonica* L.

This exotic willow grows quite commonly in the vicinity of the park. Within our area it was introduced into Cades Cove and a few other localities, invariably growing near old homesites. One of the last trees to lose its leaves in the autumn, it is the first to leaf out in the spring; over a 17-year period during which time records were maintained, the weeping willow became green with new leaves as early as January in 2 years, February in 11 years, March in 3 years, and April in 1 year (Gatlinburg).

3. COASTAL PLAIN WILLOW*—*Salix caroliniana* Michx.

This shrub (or tree) appears to be uncommon or rare in the park, occurring "in gravel bars and on stream banks at lower elevations" [Sharp, 1942a]. There is a specimen in the park's herbarium from Happy Valley, near the western boundary, at 900 ft. altitude.

4. SMALL PUSSY WILLOW†—*Salix humilis* Marsh.

This shrubby low-growing willow is uncommon in the park where it ranges from 2800 to 5800 ft. It occurs on Gregory Bald, Andrews Bald, along the Appalachian Trail in parts of the western

---

* In 1937 C. R. Ball, the noted authority on *Salix,* determined a specimen of one of our willows as *Salix longipes* var. *wardii.* According to Little [1953], this now goes by the name of *S. caroliniana.*

† Sharp et al. [1960] are of the opinion that only the variety *rigidiuscula* Anderss. occurs in the park.

half of the park, on the summit of High Rocks, along Caldwell Fork in Cataloochee, and probably elsewhere. Dates of flowering have ranged from as early as March 23 (1938, at 2800 ft.) to as late as May 29 (1939, at 5800 ft.).

5. DWARF GRAY WILLOW*—*Salix humilis var. microphylla* (Anderss.) Fern.

This willow has the appearance of a diminutive form of the small pussy willow. In the park it appears to be restricted to Parson Bald, Gregory Bald, and Spruce Mountain where it grows near the summits, ranging from 4700 to 5500 ft. Its time of flowering is May and June.

6. BLACK WILLOW—*Salix nigra* Marsh.

This, the most common willow in the park, ranges from the lowest elevations (900 ft., Abrams Creek) to 6000 ft. (vicinity of Andrews Bald). It occurs most frequently along streams at the lower altitudes. The period of flowering is from March (some years) into May, the flowers preceding the appearance of the long-pointed lanceolate leaves.

7. SILKY WILLOW—*Salix sericea* Marsh.

Ordinarily a shrub, this willow is widely distributed and quite common in the park [Jennison, 1939a]. Its altitudinal range is from the low elevations up to approximately 5300 ft. (Charlies Bunion). The silky willow, named for the glistening silky-white undersurface of the leaves, flowers in March and April. Jennison [1939a] mentions a specimen of arborescent proportions along Little River, near Elkmont. This willow has also been recorded from Cades Cove, Gatlinburg, Greenbrier, Alum Cave Parking Area, and Moores Spring (near Gregory Bald).

POPLAR—*Populus* L.

1. WHITE POPLAR—*Populus alba* L.

This exotic tree, ordinarily occurring in thickets as a result of numerous root sprouts, was introduced near a number of former homesites in the park. In the Sugarlands it persists in a few places up to an elevation of

---

* *Salix tristis* Ait. in Sharp et al. [1960].

approximately 2000 ft., while at Smokemont it grew at about 2200 ft. During fairly mild winters it has been known to come into flower in the last week of February.

2. EASTERN COTTONWOOD—*Populus deltoides* Bartr.

Although the park is within the range of this widespread tree, it is scarce or absent in the mountains. Largely on the basis of a specimen growing in Cades Cove, Sharp [1942a] states that the eastern cottonwood is "evidently introduced as it is found only near old homesites."

3. BALM-OF-GILEAD—X *Populus gileadensis* Rouleau

Gattinger [1887] stated that this tree was introduced by early settlers in East Tennessee. Galyon [1928b] is of the same opinion, adding that it sometimes escapes from cultivation. In July 1962 A. R. Shields showed H. L. Hoffman and me two places in Cades Cove where a number of these trees were growing. Shields recalls having seen the balm-of-gilead in Cataloochee. Fleetwood [1935] reported this hybrid tree from near an old house site on Mannis Branch, near Elkmont.

4. BIGTOOTH ASPEN—*Populus grandidentata* Michx.

Since the natural range of this tree appears to be outside the limits of the park, the few specimens that occur here were probably introduced. The largest tree in a small grove growing on the old Whaley homestead in Greenbrier measured 18 inches in diameter (1946). Specimens from trees growing near the Park Headquarters area and at 1600 ft. altitude in the Sugarlands were examined by Burton V. Barnes (Univ. Michigan) whose determination read "not *P. grandidentata,* possibly *P. alba* X *P. grandidentata.*"

5. LOMBARDY POPLAR—*Populus nigra* var. *italica* Muenchh.

This foreign tree is scarce in the park. It was introduced near old homesites in Cades Cove, near Park Headquarters, and probably elsewhere.

FAMILY ✦ *JUGLANDACEAE*

WALNUT—*Juglans* L.

1. BUTTERNUT—*Juglans cinerea* L.

This rather uncommon tree occurs in rich woodlands at low and middle altitudes. Shanks [1954a] lists it as a canopy tree (not dominant) in cove hardwood forests. Most butternuts grow along, or near, streams.

Ordinarily this tree is in bloom from late April to middle May.

2. MANCHU WALNUT—*Juglans mandshurica* Maxim.

In the University of Tennessee Herbarium is a specimen of this foreign tree that had been collected at an old homesite in Cades Cove in July 1938. Prior to the construction of the Sugarlands Visitor Center a Manchu walnut, along with a number of other exotic plants, was to be found growing there.

3. BLACK WALNUT—*Juglans nigra* L.

This is a more common tree than the butternut *(J. cinerea)*, but since almost all specimens grow on old homesites or in places where man was directly or indirectly involved, Jennison [1939a] believed this tree to be "adventive, but likely to become perfectly naturalized. Probably not a component of the original hardwood forests of the Park." Sharp notified me that he holds the same opinion. Shanks [1954a] listed the black walnut as a canopy tree (not dominant) in cove hardwood forests.

The highest altitude it is known to occur is at approximately 3850 ft., on the J. M. Conard place in Cataloochee. The range of its flowering is from April 7 (1945) to May 29 (1946), with normal peak of bloom in late April and early May.

HICKORY—*Carya* Nutt.

Of the seven species of hickory included here one, the pecan *(C. illinoensis),* is not indigenous and is scarce and localized; of the remaining six, only three—bitternut hickory *(C. cordiformis),* pignut hickory *(C. glabra),* and mockernut hickory *(C. tomentosa)*—occur in appreciable numbers.

"Carya is a difficult genus for identification, because it contains many minor intergrading variations in shape and size of fruits and in hairiness and other characters of foliage to which numerous names have been given, largely by C. S. Sargent and W. W. Ashe. It seems unnecessary to recognize most of these varietal names . . ." [Little, 1953].

The fruits of the various hickories are important sources of food for squirrels who begin cutting the nuts in the summer. The foliage becomes a vivid golden-yellow in the autumn.

Determinations of the University of Tennessee's specimens of *Carya* were made by Wayne E. Manning in 1948.

1. BITTERNUT HICKORY—*Carya cordiformis* (Wangenh.) K. Koch

This large tree grows in a number of cove hardwood forests but, ordinarily, is not one of the dominant species there [Jennison, 1938; Shanks, 1954a]. It is one of the entities in Whittaker's [1956] mesic tree class. According to Sharp [1942a], it grows in well-drained woods loam. Several fine mature specimens occur along the Big Locust Nature Trail at approximately 2800 ft. elevation. Whittaker's [1956] population chart shows 3150 ft. as the upper limit in these mountains.

2. PIGNUT HICKORY—*Carya glabra* (Mill.) Sweet

Shanks [1954a] called this and the mockernut hickory *(C. tomentosa)* characteristic canopy trees of the closed oak forests. According to Sharp [1942a] who regarded it as uncommon, it frequents dry sandy soils at lower elevations. Whittaker [1956] grouped it with his submesic tree class and charted it as ranging up to 4850 ft.—this makes it the hickory with the greatest altitudinal range in the park.

Flowers have been noted on the pignut hickory in middle May.

Baldwin [1948b] located a specimen of this tree (called *C. ovalis)* along Roaring Fork whose circumference measured 11 ft. 3 inches, and whose height he estimated at 125 ft.*

3. PECAN—*Carya illinoensis* (Wangenh.) K. Koch

This species was introduced into the Sugarlands Valley in 1874† when a number of young trees were set out. At least one reached a trunk diameter

* A photograph of the lower 12-15 ft. of the trunk of this tree accompanies his article [Baldwin, 1948b]. The biggest recorded individual of this species, growing in the Norris Reservoir, Tennessee, has a circumference of 13 ft. 4 inches, and a height of 145 ft. [Dixon, 1961].

† According to Earnest Ogle, Gatlinburg, the trees were set out by Perry Whaley (information recorded May 10, 1953).

of 3 ft. (1943). Some pecans still persist there (1600-2000 ft. altitude), occasionally producing some fruit.

4. SHELLBARK HICKORY—*Carya laciniosa* (Michx. f.) Loud.

Only two specimens of this hickory are known, both growing on the J. W. Owenby place, along Route 73, approximately 2 miles east of Gatlinburg. This location is within 200 yards of the park line. In 1942, when Mr. Ennis Owenby called them to my attention, he stated that the largest was estimated to be about 100 years old; in June 1963 this larger hickory measured 20½ inches in diameter and was in good condition.

5. SHAGBARK HICKORY—*Carya ovata* (Mill.) K. Koch

The only known places in the park where this well-named tree grows are along the Cherokee Orchard-Roaring Fork loop road and on Rich Mountain close to the old road from Tuckaleechee to Cades Cove. Elevations at these localities are approximately 2000-2500 ft. This rare hickory was noted to be in flower on April 30 (1937).

6. SAND HICKORY—*Carya pallida* (Ashe) Engl. & Graebn.

This, like the shellbark and shagbark hickories, is a rare tree in our area. In June 1935 Jennison noted a specimen near Deals Gap, close to the park boundary on the North Carolina (Swain County) side of the divide. He indicated that the habitat was a rich well-drained slope at about 2500 ft. altitude.

7. MOCKERNUT HICKORY—*Carya tomentosa* Nutt.

Along with the pignut hickory, this is regarded as one of the characteristic canopy trees in closed oak forests [Shanks, 1954a]. Whittaker [1956], who shows the mockernut hickory attaining an altitude of 2800 ft., groups it with trees of the submesic class. This hickory grows in Cades Cove and in places along the park's southwestern boundary.

A dense forest composed of dozens of species of trees covers the mountainsides that sweep up to the summit of the Chimney Tops. The West Fork of the Little Pigeon River and its tributaries drain this watershed.

Mountain laurel *(Kalmia latifolia)* ranges widely throughout the park, where it flowers in May and June. Ordinarily a shrub, this evergreen-leafed plant often becomes large enough to be classed as a tree.

Rosebay rhododendron *(R. maximum)* with its dark evergreen foliage is one of the most prevalent woody plants along the watercourses of the park.

Aerial photograph of Cades Cove. This flat oval-shaped valley lies north of the main Great Smoky Mountains ridge on the Tennessee side of the park. The dark spots in the foreground are cattle.

FAMILY + *CORYLACEAE*

HAZEL—*Corylus* L.

1. AMERICAN HAZELNUT—*Corylus americana* Walt.

Neither of the two hazelnuts is common in the park. The American hazelnut, however, is much more in evidence than the scarce beaked hazelnut. The former grows "in rather rich loam on moist banks and at borders of woods" [Jennison, 1939a]. In the western part of the park this shrub has been noted growing in Happy Valley, Cades Cove, Whiteoak Sink, and along Abrams and Tabcat Creeks; in the central part it occurs between Gatlinburg and Emerts Cove, in the Smokemont area, and along Mingus Creek; along the eastern boundary it has been collected at Cove Creek in Haywood County, North Carolina. The altitude at these localities ranges from approximately 900 to 2500 ft.

This is one of the earliest shrubs to bloom, flowers having been noted from January 22 (1937) to April 8 (1941). The usual flowering peak is during the latter half of February.

2. BEAKED HAZELNUT—*Corylus cornuta* Marsh.

This rare shrub has been recorded from but two localities—Happy Valley on the park's western boundary (1300 ft.) and near the summit of High Rocks (5188 ft.) on Welch Ridge.

HOPHORNBEAM—*Ostrya* Scop.

EASTERN HOPHORNBEAM*—*Ostrya virginiana* (Mill.) K. Koch

Although Shanks [1954a] lists it as one of the small trees characteristic of the cove hardwood forests and Whittaker [1956] includes it in his mesic tree class, this is an uncommon to rather scarce species over most of the park. As Sharp [1942a] has stated, the hophornbeam is most likely to occur in limestone areas, but it is not restricted to alkaline soils. This tree has been recorded from approximately 900 ft. (mouth of Abrams Creek) to

---

* In November 1962 A. J. Sharp informed me that he believed both *Ostrya virginiana* and the variety *lasia* Fern. occurred in the park.

4500 ft. (Hyatt Ridge Trail). Mature specimens grow along lower Little River (Tuckaleechee Cove), in Whiteoak Sink and Wears Cove, near Mt. Sterling Gap and elsewhere.

In our area this tree is ordinarily under 12 inches in diameter and less than 60 ft. in height. Flowers have been noted from late March (1944) to late April (1937) in Gatlinburg.

HORNBEAM—*Carpinus* L.

AMERICAN HORNBEAM; IRONWOOD—
Carpinus caroliniana Walt.

This is a common small tree along most of the watercourses in the park below the 3000 ft. altitude. Shanks [1954a] calls it one of the small trees characteristic of the cove hardwood forests, but this applies only to the lower limits of that type of forest. The American hornbeam grows near the 3000 ft. altitude above Smokemont and in Cataloochee. It is one of the species comprising Whittaker's [1956] mesic tree class.

Flowers have appeared on this hornbeam as early as March 17 (1945), but ordinarily it is at the end of March and in early April that this tree comes into bloom. Gray squirrels have been observed feeding on the small fruits in early October.

An unusually large specimen with a circumference of 7 ft. 7 inches, a height of approximately 42 ft., and a spread of about 30 ft.* grows in the Big Creek area of the park approximately ¼ mile upstream from the campground and about 100 ft. from the river. The altitude there is near 1650 ft.

BIRCH—*Betula* L.

1. YELLOW BIRCH†—*Betula alleghaniensis* Britton
On undisturbed mountainsides at elevations of 3500-4500 ft.,

---

* A specimen measuring 5 ft. 7 inches in circumference, growing in Princess Anne, Md., is listed as the largest on record by Dixon [1961].

† A number of botanists (Sargent, 1933; Coker and Totten, 1934; Fernald, 1950) refer to this tree as *B. lutea* Michx. Whittaker [1956] includes both *B. alleghaniensis* and *B. lutea* from the park, the latter from elevations around 4500 ft. and the former below that altitude. Little [1953], the authority followed in the present report, rejects *B. lutea*.

especially throughout the eastern two-thirds of the park, this is one of the most abundant trees in the area. Shanks [1954a] lists it as a characteristic canopy tree in cove hardwood, hemlock, northern hardwood, and spruce-fir forests. From about 1200 ft. (Little River) it ranges up to at least 6400 ft. on some of the highest mountains. A large proportion of the mature trees one sees along the Tennessee side of the main transmountain road between the tunnel overpass and Walker Prong (3500-4500 ft.) are yellow birches.

Ordinarily this tree is in flower during the first half of May, but it has come into bloom as early as April 9 (1946). At high altitudes, flowers have been noted as late as June 5 (1940).

One of the park's "champion" trees is a yellow birch measuring 14 ft. 1 inch in circumference [Dixon, 1961].* This was reported by Baldwin [1948b] who discovered it on False Gap Prong in the Greenbrier area. Many specimens of yellow birch exceed 10 ft. in circumference. Burls of varying sizes often develop in the larger trees, and, as both Campbell [1936] and Cain [1940] have reported, this tree often develops a stilt- or prop-growth as the result of the seed having germinated on the top of a rotting stump or log. Cain [ibid.] also calls attention to this tree's habit of sending down aerial roots along the outside of the trunk. Sharp [1957] mentions several kinds of trees and shrubs that may be epiphytic on yellow birch in the Great Smoky Mountains, at times growing in crotches or crevices 40 to 50 ft. above the ground.

Ordinarily it is during the second and third weeks in October that the leaves of this species assume a rich yellow-gold color, and by the end of that month these trees are quite defoliated.

2. SWEET BIRCH—*Betula lenta* L.

This birch, distinguished by its cherry-like bark, is listed by Shanks [1954a] as being a characteristic canopy tree in closed oak and hemlock forests, a non-dominant species in cove hardwoods, and occurring occasionally in spruce-fir forests. Whittaker [1956], who groups it with trees in his "submesic" class, indicates that it ranges upward to 5600 ft. in red oak-chestnut forests, but ordi-

---

* If it is still standing, a yellow birch on the summit of White Top Mountain, Virginia, measuring 7 ft. 3 inches in diameter, is larger [Coker and Totten, 1934].

narily it is scarce or absent above 4500 ft. It grows near 1000 ft. altitude along the lower part of Panther Creek and near 4900 ft. on Gregory Bald and along the Appalachian Trail near Mt. Cammerer. Over the park as a whole it is not nearly as abundant as the yellow birch.

The sweet birch displays its flowering tassels in April, particularly during the latter half of that month.

Like the leaves of other birches, those of the sweet birch become a rich golden color in October.

A specimen found in the park and reported by Baldwin [Dixon, 1961] is listed as the largest on record; it measures 10 ft. 10 inches in circumference. This tree grows on the south side of the trail to Ramsey Cascades, along Ramsey Prong.*

3. RIVER BIRCH—*Betula nigra* L.

Of the three native birches, this is by far the scarcest. It occurs quite sparingly along some of our low altitude watercourses, rarely penetrating above the 2000 ft. altitude. However, I discovered a specimen 3 to 4 inches in diameter growing along Little River, above Elkmont, at 3000 ft.; this probably marks its highest occurrence in these mountains. The river birch is to be found in Cades Cove, and along Abrams Creek, Eagle Creek, and the lower parts of Little River. Along the Little Tennessee and Tuckasegee Rivers, near the park, it becomes quite common.

The flowers of this birch appear in April.

The bark of the river birch usually assumes a shaggy appearance due to the exfoliation of the thin buff or pinkish-buff layers.

ALDER—*Alnus* B. Ehrh.

HAZEL ALDER—*Alnus serrulata* (Ait.) Willd.

Only one species of alder grows in the park. Ordinarily its size and growth-habits are shrub-like, but occasional specimens are found that may qualify for the designation of tree. The hazel alder occurs commonly along streamsides and in moist situations up to an altitude of approximately 3000 ft. It is one of the earliest

---

* According to Coker and Totten [1934], this birch has been reported up to a diameter of 6 ft., but no locality is given. Such a specimen would be considerably larger than the one mentioned above.

of all our woody plants to come into bloom; during the 26-year period beginning in 1937, there were 8 years when alders were in flower before the close of January (earliest: January 14, 1950). In most years the tassels of flowers are unfurled in February.

In all probability, specimens bigger than the 22-inch circumference alder I measured along Laurel Creek (1400 ft.) will be found in our area.

FAMILY ✦ *FAGACEAE*

BEECH—*Fagus* L.

AMERICAN BEECH—*Fagus grandifolia* Ehrh.

This tree is a dominant species in cove hardwood, northern hardwood, and in hemlock forests, and it is of occasional occurrence in the spruce-fir forests [Shanks, 1954a]. While it may occur in pure stands in some of the beech gaps along the high state-line ridge, it is absent from extensive areas where oaks and pines prevail. Whittaker [1956], who groups the beech in a "mesic" tree class, follows Camp [1951] in distinguishing three kinds of beeches: "At lowest elevations in cove forests and cove forest transition the 'white' beech, of primarily southern distribution, may be recognized. About 1000 ft. above the upper limit of the white beech, other, 'red' beeches appear in the upper cove forests (3500-4500 ft.). The third beech type, the 'gray,' appears in the beech gap forests of still higher elevations, above 4500 ft."

The American beech approaches its uppermost range near the southwest margin of Andrews Bald (5800 ft.) and at 6000 ft. along the Appalachian Trail approximately 1½ miles east of Newfound Gap.

Beech gaps, such as the one on the main divide between Newfound Gap and Indian Gap, occur in a number of places at high altitudes. As the result of his study on the persistence of beech in such places, Russell [1953] concluded "that the ability of the beech to stand great wind damage in the gaps was a major factor in its continued survival." Ganier [1956], in his study of the nesting sites of the black-throated blue warbler in the nearby

Unicoi Mountains, believes that "The presence of a few cattle along the crestline forest has resulted in the young beech trees being kept cropped a few feet from the ground causing them to be quite thick." Whether this was also a factor in the Great Smoky Mountains is not known, although it is known that large numbers of cattle had free range in the higher elevations in pre-park days. As mentioned in the account of the red spruce in this report, Whittaker [1956] is of the belief that "The spruce forests should have been moving southwest along the ridge from Clingmans Dome in the 4,000 years since the peak of the xerothermic period [Flint, 1947], but are perhaps retarded or halted by the extensive beech forests of Double Spring Gap."

The flowers of American beech were noted from April 11 to May 22 (1953). On occasion in the high altitudes, late frosts kill the young leaves over wide areas. The tenacity of the mature leaves is considerable; I have noted them on young trees near Newfound Gap as late as May 11 (1950), thus persisting for approximately 11 months.

CHESTNUT—*Castanea* Mill.

1. AMERICAN CHESTNUT—*Castanea dentata* (Marsh.) Borkh.

This ill-fated tree was one of the largest, most abundant, and most valuable species in the eastern forests. Sprouts still continue to grow from some of the old stumps, persisting for a few years before yielding to the ravages of the oriental blight that has eliminated the American chestnut from these mountains. Here and there a few of the dead trees still stand reminding us of a tragedy the like of which has no counterpart in forest annals.*

The American chestnut was a dominant species in cove hardwoods, closed oak forests, and in open oak and pine stands [Shanks, 1954a]. Whittaker [1956] who placed it in his subxeric tree class indicated its occurrence to 6000 ft., but ordinarily the uppermost penetration was between 5000 and 5500 ft. From there it extended down to the lowest altitudes. Ayres and Ashe [1905] in their report of more than a half-century ago indicated

* Readers are referred to the chapter entitled "Farewell to the Chestnut," in the *Great Smoky Mountains National Park Natural History Handbook* [Stupka, 1960].

that chestnut made up 30 percent of the forest cover in the Cades Cove district and 40 percent on the ridges in the Cataloochee district.

Chestnut would come into flower in June at the lower and middle altitudes and in July above 3500 ft. I have noted blossoms on a weakening tree at Alum Cave Bluffs (4900 ft.) as late as August 5 (1952). Along with his monthly report for November 1935, Jennison [1935b] submitted a list entitled "Trees of the Great Smoky Mountains National Park" in which he makes the following observation: "Most of our chestnut is gone but on the sheltered, moist north-east slope of the State Line Ridge between Gregorys Mt. and Ekaneetlee Gap one may still see a chestnut-peawood [silverbell] forest covering hundreds of acres, little damaged by the blight as yet, so little in fact as to appear as though snow had fallen on it when the trees were in bloom last summer."

Harvey Broome—lawyer, hiking enthusiast, and prominent conservationist from Knoxville, Tennessee—informed me (1945) that the chestnut trees in the Cherokee Orchard area showed no blight effects in 1922, but that some dead snags appeared there a few years later. From their study of the natural replacement of chestnut by other species in Great Smoky Mountains National Park, Woods and Shanks [1957] conclude that the arrival date of the fungus was approximately 1925-26. Jennison, in a letter dated June 17, 1937, notified Dr. Arthur H. Graves of the Brooklyn Botanic Garden that "There are few if any trees in this section which are wholly free of blight disease. However, some at higher elevations are still in 'pretty good shape.' " F. H. Miller [1938], park forester, estimated that by 1938 "fully 85% of the chestnut in the Park has been killed or is affected by the blight." In the same year he reported [ibid.] that "Only one section of the Park was noted for the low count of dead or dying chestnuts, namely up Caldwell Fork of Cataloochee Creek Watershed where trees 4 feet in diameter are not uncommon."

Only the yellow-poplar would attain the great size of the American chestnut; these two trees were the monarchs of the Southern Appalachian forests. Buckley [1859], more than a century ago, reported measuring a chestnut tree somewhere on the western slope of the Great Smoky Mountains in Tennessee that was "thirty-three feet in circumference at four feet from the ground. It is a noble living specimen apparently sound, and of nearly a uniform diameter upwards for forty or fifty feet." The photograph of a slightly larger tree from within the park area appeared in the *Journal of Heredity* for September 1915 (p. 412). Entitled "Chestnut Tree from North Carolina," the caption reads "After the California oak, the largest nut-bearing tree discovered by the association is this chestnut *(Castanea dentata)* 3 miles from Crestmont, North Carolina. It stands on the main range of the Big Smoky Mountains, dividing Tennessee and North Carolina; the altitude of the location where the tree grows is about 2900 feet. This specimen is about 75 feet high; its girth at 7 feet from the ground is 33 feet 4 inches." (The location of Crestmont is on Big Creek, about 1 mile inside the park boundary.)

Dr. A. J. Sharp informed me that in Greenbrier, in 1934, he measured a chestnut stump that was 13 ft. the long way across at ankle-high. Miller [1938] refers to a specimen 9 ft. in diameter that had been "cut for honey a few years ago"; its location is not given. Wilbur L. Savage who was then serving as forester in Great Smoky Mountains National Park reported to me that on Dunn's Creek on September 7, 1943, he measured a fallen dead chestnut that was 9 ft. 6 inches in diameter; at 65 feet from the ground this tree measured 4 ft. 8 inches in diameter. In May 1943 Savage came upon a chestnut measuring 27 ft. 3 inches in circumference, the location being along the Baxter Creek Trail to Mt. Sterling.

A good crop of fruits was noted on the American chestnuts on Welch Bald in September 1938. It is unlikely that this fine tree bore much of a harvest after the 1940's. The trees at the highest elevations were ordinarily the last to succumb to the relentless disease.

According to Woods and Shanks [1959] who made a study of the natural replacement of chestnut by other trees in the Great Smoky Mountains National Park, five species of oaks together made up 41 percent of the various kinds of trees involved. Of these the chestnut oak *(Q. prinus)* and the northern red oak *(Q. rubra)* had the greatest proportion of all replacements, comprising 17 percent and 16 percent, respectively.

2. CHINESE CHESTNUT—*Castanea mollissima* Bl.

One of these exotic trees grows along the loop road in Cades Cove. It is possible that this species was introduced in a few other locations.

3. ALLEGHENY CHINKAPIN—*Castanea pumila* Mill.

This plant is either very rare or absent. Jennison [1939a] stated that it grew "in well drained sandy soil at lower elevations. . . . The few specimens of this species known to occur in the Park have been all but destroyed by the chestnut bark disease. One standing near Smokemont was represented (1937) by a vigorous sprout." This may have been the specimen H. C. Wilburn measured (15 inches in diameter) and photographed* March 4, 1937, and regarding which he stated, "It served many years as a 'corner tree' to certain of the Bradley tracts at and near Smokemont."

In the University of Tennessee Herbarium are two specimens —one from Smokemont (probably the above-mentioned sprout of an old tree that bore fruits on July 30, 1937) and the other from Mt. Sterling, North Carolina, on the park's northeastern boundary (collected August 2, 1902, by Albert Ruth).

OAK—*Quercus* L.

1. WHITE OAK—*Quercus alba* L.

This is one of the most abundant of the 10 species of oaks in the park, being a dominant component of both the closed oak forests and the open oak and pine stands [Shanks, 1954a]. Whittaker [1956] who places it in the subxeric tree class believes that two population-types are represented in the park "separated by an elevation gap of 1000 ft. or more." White oaks occur commonly enough throughout the lower altitudes; in the higher elevations they may reach to 5900 ft. [ibid.].

Flowers have been noted on this tree as early as April 11 (1945) and as late as May 26 (1949)—the latter date at 4750 ft. The acorns appear to be a favorite food item with bears and other forms of wildlife.

Buckley [1859] stated that "On Jonathans Creek there is a white oak *(Quercus alba)* nineteen feet in circumference at three feet from the ground." That locality lies just outside the park near its southeastern boundary.

* In photographic files, Great Smoky Mountains National Park.

2. SWAMP WHITE OAK—*Quercus bicolor* Willd.

The only place in the park where this oak is known to occur is on a rock-strewn hillside close to the loop road in Cades Cove. Three trees comprise the "grove," one of which consists of four main trunks that branch at or near the ground; two additional single specimens stand 30-40 feet away. All these trees have trunks measuring approximately a foot in diameter (1963). The location is 1.4 miles beyond the Cades Cove Orientation Shelter; the quadruple specimen grows 10 feet to the right (north) of the road and its branches form an archway over it. In the immediate vicinity stands a towering black walnut *(Juglans nigra);* a few shingle oaks *(Quercus imbricaria),* red mulberries *(Morus rubra),* flowering dogwoods *(Cornus florida),* yellow-poplars *(Liriodendron tulipifera),* and black locusts *(Robinia pseudoacacia)* grow close by.

3. SCARLET OAK—*Quercus coccinea* Muenchh.

This is a common tree below 3500 ft. on fairly dry slopes and ridges, especially where pines predominate. Jennison [1939a] called it, "A frequent component of the yellow pine-hardwood and oak-chestnut forest types," while Shanks [1954a] stated that it was dominant in open oak and pine stands. Whittaker [1956], who gives 4200 ft. as the highest altitude for the scarlet oak, places it in his xeric tree class along with three species of pines *(Pinus virginiana, P. rigida, P. pungens)* and the blackjack oak *(Quercus marilandica).*

The flowers of the scarlet oak appear in April. The brilliant glossy-red leaves in late October and early November make this the most attractive of the oaks in the autumn; this beauty is enhanced by the color contrast with evergreen pines that are often associated with it. The common name of no other kind of oak is more appropriate than this one.

A specimen of scarlet oak measuring 11 ft. in circumference grows in the campground area at Smokemont.

4. SOUTHERN RED OAK*—*Quercus falcata* Michx.

This tree, like the scarlet oak, grows in dry woodlands at the lower altitudes, but it is not nearly as common as the scarlet oak. Jennison [1939a] stated that the southern red oak was "Most frequent in oak-chestnut forest types; less frequent in yellow pine-hardwood types." It grows in Cades Cove, Emerts Cove, Wears Cove, and elsewhere, usually at altitudes below 2000 ft.

Flowers have been noted in April. The leaves of this oak are exceptionally variable, but the frequent presence of a long middle "finger" will help to identify this tree when in leaf.

5. SHINGLE OAK—*Quercus imbricaria* Michx.

This is one of the rather scarce oaks in the park. It grows "in rich sandy loams at lower elevations" [Jennison, 1939a], usually along streams. Specimens occur in Cades Cove on the Tennessee side of the park and along Deep Creek, Indian Creek, Couches Creek, and the Oconaluftee River (near the Visitor Center) on the North Carolina side. All localities are below 2100 ft. in altitude.

Like most of the other oaks, this species flowers in April and May when its new leaves are quite small. It is our only oak with entire leaves, these being oblong-lanceolate, thick, and of a dark glossy-green color.

6. BLACKJACK OAK—*Quercus marilandica* Muenchh.

This is an uncommon small tree in the park where it may occur in poor sandy soils usually below 2000 ft. Whittaker [1956], who gives 2400 ft. as its highest occurrence in these mountains, includes the blackjack oak in his xeric tree class along with the scarlet oak *(Q. coccinea)* and three pines *(Pinus virginiana, P. rigida,* and *P. pungens).* It grows on Rich Mountain, in Cades Cove, near Laurel Lake, on Mt. Harrison, near Deals Gap, and above Fontana Reservoir.

"The leaves are narrow at the base and very broad in their apical third" [Sharp, 1942a]. They may resemble the leaves of the black oak *(Q. velutina),* but the leaves of the blackjack oak

---

* According to information received from A. J. Sharp (1963), there are at least two hybrids in the Cades Cove area. One is *Quercus* X *anceps* Palmer, a hybrid of *Q. falcata* and *Q. imbricaria;* the other is *Quercus* X *willdenowiana* (Dippel) Zabel, a hybrid of *Q. falcata* and *Q. velutina.*

are dark green, thick, and leathery with a rusty-brown scurfiness on the undersurface.

7. CHESTNUT OAK—*Quercus prinus* L.

This is a common tree in rocky soils at low and middle altitudes. Shanks [1954a] lists it as a dominant species both in closed oak forests and in open oak and pine stands. Whittaker [1956] who includes the chestnut oak in his subxeric tree class gives 4700 ft. as the highest altitude it attains; ordinarily this tree becomes rather scarce above 4000 ft.

Flowers have been noted in April and May. Large chestnut oaks grow along the lower part of the trail to Ramsey Cascades and elsewhere. In the Cades Cove area, about ½ mile below the falls on Mill Creek, is a tree 16 ft. 9 inches in circumference. According to Ayres and Ashe [1905], 30 percent of the forest covering the ridges in the Cataloochee district consisted of chestnut oak; in Cades Cove this amounted to 20 percent.

Lambert [1958] remarked that during the logging era the chestnut oak was the largest contributor to the tanbark industry in the Southern Appalachian area. The study by Woods and Shanks [1957] on the natural replacement of chestnut by other species in Great Smoky Mountains National Park revealed that chestnut oak, with 17 percent of all replacements, led all the other trees.

8. NORTHERN RED OAK—*Quercus rubra* L.

This large tree is a dominant species in cove hardwoods, in hemlock forests, and in closed oak forests [Shanks, 1954a]. Whittaker [1956], who places the northern red oak in his submesic tree class, is of the opinion that two populations—one at high and the other at low altitudes—are represented and that their characteristics change gradually with change in altitude.

This oak grows along the margins of a number of the grass balds in the park (Andrews, Gregory, Parson, and Spence Field). Whittaker's [1956] population chart shows 6000 ft. as the highest altitude it attains. Fine mature specimens grow along the Buckeye and Junglebrook Nature Trails. Occasional trees occur below 1500 ft., but at that elevation it is rather scarce.

Flowers appear in April and May. It was in good flower

May 20, 1943, at Pin Oak Gap (4428 ft.) and May 26, 1949, near Devils Tater Patch (4750 ft.).

This big-fruited oak attains considerable size, and a number of specimens are known whose circumference ranges from 15 to 19 ft. The biggest, a specimen on Rowans Creek (Cades Cove) that died about 1958, measures 19 ft. 4 inches in girth.

Woods and Shanks [1957] in their study of the natural re-placement of chestnut by other species in Great Smoky Mountains National Park discovered that the northern red oak with 16 percent of all replacements was second only to the chestnut oak (17 per-cent) in this regard.

9. POST OAK—*Quercus stellata* Wangenh.

Although common enough in Cades Cove, the post oak is rather uncommon throughout the rest of the park. Jennison [1939a] regarded it as "An infrequent component of the oak-chestnut forest type." Sharp [1942a] who says it grows in "dry soils of exposed ridges at lower elevations" recorded a specimen from 4500 ft. on Greenbrier Pinnacle, but ordinarily it is confined to elevations below 3000 ft.

The flowers of the post oak appear mostly in April. The broad, usually squarish, lobes tend to give the leaf the appearance of a cross.

10. BLACK OAK*—*Quercus velutina* Lam.

This tree is dominant in both the closed oak forests and open oak and pine stands [Shanks, 1954a]. It grows in well-drained soils from the lowest altitudes to 4800 ft. [Whittaker, 1956].

Flowers appear in April or May. The leaves, variable in shape, are usually quite broad, leathery, and the lobes are terminated by bristle tips. The bark of mature trees is black in color.

* See footnote under *Quercus falcata* re: *Q.* X *willdenowiana,* said to be a hybrid between *Q. velutina* and *Q. falcata.*

FAMILY ✦ *ULMACEAE*

ELM—*Ulmus* L.

1. WINGED ELM—*Ulmus alata* Michx.

Along with sycamore and American hornbeam, the winged elm is a common streamside tree at the lower altitudes. It appears to grow only at elevations below 2000 ft.

This elm is usually the earliest of our native trees to come into flower. During the 5-year period beginning in 1949, winged elms came into bloom on four of the five Januaries; earliest was on January 15, 1950. Ordinarily, however, this tree begins its period of flowering in February. If the winter has been prolonged and quite cold, the bloom may not appear until March.

Some of these trees appear to lack the corky wings on the twigs, a character that gives this species its name.

2. AMERICAN ELM—*Ulmus americana* L.

This appears to be an uncommon tree in the park where it may occur "in moist soils at lowest elevations only" [Sharp, 1942a]. Specimens have been recorded from Laurel Creek, Whiteoak Sink, Elkmont, and the Park Headquarters area—all on the Tennessee side of the park.* If the Elkmont record represents an indigenous tree, the highest altitude for the species here would be approximately 2200 ft.; otherwise 1750 ft. would mark its upper limits. It appears that the fatal Dutch elm disease that has killed so many American elms in eastern Tennessee and elsewhere will, in all probability, eliminate this tree from the park.

Flowers have been noted as early as January 26 (1937) and as late as March 29 (1941).

3. SLIPPERY ELM—*Ulmus rubra* Muhl.

This is an uncommon tree "in moist soils at lower elevations" [Sharp, 1942a]. It occurs near the Cherokee Indian Reservation at Ela, North Carolina, and on the Tennessee side of the park in Cades Cove, Whiteoak Sink, Little River (near the Sinks), and the

* Coker and Totten [1934] and Munns [1938] indicate that the American elm does not occur in the western third of North Carolina.

Park Headquarters area. It appears to be confined to altitudes under 2000 ft.

The flowers of the slippery elm have been noted as early as January 25 (1949) and as late as March 13 (1940). Most of the records of flowering are in February.

HACKBERRY—*Celtis* L.

Little [1953] writes that *"Celtis* is a difficult genus with intergrading forms as well as ecological variations. Since the species are not well defined, it seems unnecessary to distinguish minor variations as varieties. Considerable study of the genus, both in field and herbarium, is needed." Coker and Totten [1934], Fernald [1950], Braun [1961], and others are essentially in agreement with these comments.

In addition to the two species included here, *C. occidentalis* L. has been listed for the park [Sharp et al., 1960; Shanks, 1961], but specimens appear to be lacking.

1. SUGARBERRY*—*Celtis laevigata* Willd.

For the most part, this is a rare tree in the park. Sharp [1942a] states that it occurs "in alluvial soils at lower elevations." Specimens have been noted along lower Abrams Creek, along Hesse Creek, in the Park Headquarters area, and in the Sugarlands. At the last-mentioned locality it grows along the West Prong of the Little Pigeon River at approximately 1800 ft.—the highest altitude it appears to attain.

In the Park Headquarters area flowers have been noted from April 9 (1946) to May 1 (1951).

2. GEORGIA HACKBERRY—*Celtis tenuifolia* Nutt.

This is a shrub of rare occurrence that appears to be restricted to limestone outcrops at low altitudes in the western end of the park. It has been observed near Mill Creek (Cades Cove), along Tabcat Creek, and in Whiteoak Sink.

* Shanks [1961] includes the narrow-leafed extreme variety *smallii* (Beadle) Sarg. in Sevier County, Tennessee.

FAMILY ✦ *MORACEAE*

MULBERRY—*Morus* L.

1. WHITE MULBERRY—*Morus alba* L.

For many years one of these foreign trees grew near a residence just a few hundred yards upstream from the Deep Creek Campground. Another was reported growing near a homesite along the road from Gatlinburg to Cherokee Orchard.

2. RED MULBERRY—*Morus rubra* L.

Although mostly rather uncommon, this tree occurs in a number of low-altitude localities—Happy Valley, lower Abrams Creek, Cades Cove, Laurel Creek, Wears Cove, Elkmont, Greenbrier, Ravensford, and elsewhere. It reaches its highest elevation at about 2200 ft.

Flowers have been noted from April 16 (1945) to May 20 (1943). This native mulberry has large leaves that are very rough and seldom lobed. (The exotic white mulberry has smooth shining leaves with asymmetrical lobes.)

OSAGE-ORANGE—*Maclura* Nutt.

OSAGE-ORANGE—*Maclura pomifera* (Raf.) Schneid.

In all probability, this tree was introduced into the area by early white inhabitants. It persists and continues to bear its large greenish fruits in the immediate vicinity of former homesites in the Sugarlands and along Parsons Branch (Tennessee side of the park), and at Smokemont and in the Visitor Center area along the Oconaluftee River (North Carolina side).

Flowers have been noted from May 30 (1944) to June 25 (1935).

FAMILY ✦ *SANTALACEAE*

OILNUT—*Pyrularia* Michx.

OILNUT—*Pyrularia pubera* Michx.

This Southern Appalachian endemic is listed by Shanks [1954a] as a dominant shrub in cove hardwoods, closed oak forests, and in open oak and pine stands. It occurs at low and middle altitudes up to approximately 4000 ft. The oilnut comprises one of the species in Whittaker's [1956] submesic shrub union.

Flowers have been noted on this common shrub from April 11 (1945) to June 8 (1940). The usual period of flowering is during the first half of May.

Fernald [1950] calls attention to the poisonous oil in the fruit* and in other parts of this plant; he likewise mentions that it is parasitic upon roots of deciduous trees and shrubs. Totten [1937], however, remarks that "It is found associated with quite a variety of plants and its rather elaborate underground runner system, connecting what at first appears to be separate plants, shows little proof that it is sometimes a parasite, the question is again raised as to whether it is an obligate parasite or can also exist independently."

FAMILY + *LORANTHACEAE*

FALSE MISTLETOE—*Phoradendron* Nutt.

AMERICAN MISTLETOE—*Phoradendron flavescens* (Pursh) Nutt.

This parasitic evergreen is rather uncommon and is restricted to altitudes below 2500 ft. It usually grows high up on various kinds of oaks, the blackgum *(Nyssa sylvatica),* and on other deciduous trees.

The inconspicuous flowers appear in the autumn. According to Braun [1961], the whitish sticky fruits are poisonous.

---

* "So oily that it will burn like a candle if a wick be drawn through it" [Buckley, 1859].

FAMILY ✦ *ARISTOLOCHIACEAE*

BIRTHWORT—*Aristolochia* L.

DUTCHMAN'S-PIPE—*Aristolochia durior* Hill

This rope-like woody vine with its big valentine-heart leaves is one of the most characteristic plants of the cove hardwood forests. However, it is not restricted to that type of forest. It has been recorded from altitudes as low as 950 ft. on Abrams Creek up to 4800 ft. on Spruce Mountain.

May is the usual month for its flowering. The earliest blossom was noted April 16 (1945); the latest, June 11 (1936).

This vine has been observed reaching to a height of fully 100 ft. in large forest trees. Such specimens may attain a diameter of 4 inches.

FAMILY ✦ *RANUNCULACEAE*

CLEMATIS—*Clematis* L.

1. JAPANESE CLEMATIS—*Clematis paniculata* Thunb.

This exotic plant was introduced in Elkmont and grows in some gardens close to the park. "Flowers resemble those of the virgin's-bower but the leaves . . . are more glossy and coriaceous and less deeply cut" [Sharp, 1942b].

2. LEATHER-FLOWER—*Clematis viorna* L.

This woody vine is quite scarce in the park where specimens have been noted from the lowest altitudes (mouth of Abrams Creek) to 5000 ft. [Jennison, 1939a]. It has been found in well-drained soils [Sharp, 1942b] at Mt. Sterling Gap, Wears Cove, and along the trail from Cosby to Low Gap.

Flowers have been recorded from May 8 (1937) to July 18 (1935).

3. VIRGIN'S-BOWER—*Clematis virginiana* L.

This common vine is widespread over much of the low-altitude areas where it grows in open situations. It is rare or absent above 2500 ft. The first half of August marks the peak of its flowering period, but flowers may appear in middle July, and some persist until the end of August.

Following its period of blossoming, the silvery-gray plumes that adorn the seeds are quite showy, and these may persist throughout the autumn.

The leaves of this vine occur in threes, but a careful observer should experience no difficulty in distinguishing this plant from the poison ivy *(Rhus radicans)* whose leaves are thicker, larger (usually), and lacking the pronounced venation.

SHRUB-YELLOWROOT—
Xanthorhiza Marsh.

YELLOWROOT—*Xanthorhiza simplicissima* Marsh.

This is a fairly common streamside shrub from the lowest altitudes up to approximately 2500 ft. Occasionally it forms a dense ground cover. Flowers have been noted as early as February 18 (1938) and as late as April 24 (1942); ordinarily it is in blossom by late March.

Localities where it occurs include Cades Cove, Rich Mountain, Abrams Creek, and the Park Headquarters area on the Tennessee side; and Cataloochee Creek, Oconaluftee River, Deep Creek, and Bryson Place on the North Carolina side.

The bright yellow color under the bark of the roots gives this plant its name.

FAMILY ✦ *BERBERIDACEAE*

BARBERRY—*Berberis* L.

COMMON BARBERRY—*Berberis vulgaris* L.

This foreign shrub, the alternate host for the wheat rust fungus, is rare in the park. It was introduced at an old homesite along the Chestnut Flats road, near Cades Cove, many years ago [Jennison, 1939a].

FAMILY ✦ *MENISPERMACEAE*

CORALBEADS—*Cocculus* DC.

RED-BERRIED MOONSEED—*Cocculus carolinus* (L.) DC.

This semi-woody vine is a rare plant in the park where it occurs only at the lowest elevations (Abrams Creek) near the western limits of the area. The greenish flowers appear in summer. In the fall the clusters of bright scarlet fruits are quite showy.

It should be looked for "along fence rows and in thickets" [Sharp, 1942a].

MOONSEED—*Menispermum* L.

YELLOW PARILLA—*Menispermum canadense* L.

This woody vine is of rare occurrence in our area. Within the park it has been noted only near the Abrams Creek Ranger Station, close to the western boundary. It also grows at Ela, North Carolina, near the Cherokee Indian Reservation. These localities are below 2000 ft. in altitude.

The inconspicuous greenish-white flowers appear in June or July [Grimm, 1957]. The clusters of dark-colored fruits resemble small bunches of grapes. Jennison [1939a] gives its habitat as "Moist rich woods-loam, along streams and bordering woods."

FAMILY ✦ *MAGNOLIACEAE*

MAGNOLIA—*Magnolia* L.

1. CUCUMBERTREE—*Magnolia acuminata* L.

Although this tree is fairly common in some areas, over the park as a whole its status is more accurately described as rather uncommon. Shanks [1954a] lists it as one of the dominant trees in cove hardwood forests, while Whittaker [1956] includes all three of our native magnolias in his mesic tree class. The cucumbertree occurs from the lower altitudes up to approximately 5000 ft.

(Thomas Ridge; edge of Spence Field Bald). Flowers have been noted as early as April 16 (1945) and as late as May 20 (1947).

The cucumbertree has small asymmetrical fruits in contrast to the larger symmetrical fruits of *Magnolia fraseri* and *M. tripetala*. The bark of mature trees resembles that of similar-sized yellow-poplars *(Liriodendron tulipifera)*—quite unlike the smooth gray bark of *Magnolia fraseri* and *M. tripetala*. Also the leaves of the cucumbertree are smaller than the leaves of the other two native magnolias.

In the Greenbrier district, at 3050 ft. elevation on Kalanu Prong, stands a venerable cucumbertree that measures 18 ft. 4 inches in circumference—a record size for this species [Dixon, 1961]. In August 1959 only two of its large limbs were still alive.

A variety known as the yellow cucumbertree, *Magnolia acuminata* var. *cordata* (Michx.) Sarg.,* grows in the park and vicinity at a number of localities. Specimens up to 12 inches in diameter and 50 ft. in height have been noted. The flowers are a bright canary yellow. This tree grows in the following places:

1. Along the transmountain road, Tennessee side, at approximately 3050 ft. (in flower April 22, 1946, and May 14, 1947).
2. Along the Two-Mile Branch Trail, near Park Headquarters, at 2300-2400 ft. (in flower April 14, 1946, and April 29, 1944).
3. Along the Gregory Ridge Trail, near Cades Cove, at 2700 ft. (in flower May 5, 1949).
4. Along the Laurel Falls Trail to Cove Mountain, at 2700 ft., just beyond Laurel Falls (in flower May 8, 1950).
5. Along the Ramsey Cascades Trail at 3150 ft. (in flower May 19, 1950).
6. Along the Huskey Gap Trail from the Sugarlands (in flower May 2, 1961).
7. Along the Wears Cove Road near Cove Creek Cascades, outside the park, at approximately 1300 ft. (in flower April 16, 1945).

2. FRASER MAGNOLIA—*Magnolia fraseri* Walt.

This tree of the Southern Appalachian Mountains is fairly common

---

* Hardin [1954], in a report on this plant, comes to the conclusion that "The flower color is the only character found to separate these yellow-flowered forms with glabrous twigs. It is necessary, therefore, to reduce the original variety to a forma *aurea*."

below 5000 ft. except in rather dry situations. Shanks [1954a] listed it as a characteristic small tree of the cove hardwoods and a dominant canopy tree in hemlock forests and in northern hardwood forests. Whittaker [1956] places it in his mesic tree class. It ranges in altitude from 1150 ft. (Abrams Creek) to 5150 ft. (near Spruce Mountain).

The large creamy-white flowers ordinarily appear in late April or early May. However these have been noted as early as the end of March (1938, 1945). Occasionally flowers may remain on the trees into June (June 13, 1940) at the higher altitudes. The fruits begin to turn color in July and are bright red by late July and through August.

A record-size Fraser magnolia, 9 ft. 3 inches in circumference, grows near Anthony Creek along the trail to Russell Field at about 2700 ft. altitude. Within the park are a number of specimens over 8 ft. in circumference and at least 80 ft. in height.

This tree is readily identified by its smooth light gray bark, its large, eared leaves arranged in superficial whorls, and its large cream-colored flowers which later give way to the cucumber-like fruits. Quite frequently, several stems or sprouts arise from the base of this tree.

3. SOUTHERN MAGNOLIA—*Magnolia grandiflora* L.

In the University of Tennessee Herbarium is a specimen of this introduced tree that came from the Park Headquarters area where it had been planted many years ago. Although the southern magnolia is no longer present there, it may possibly persist in the vicinity of an old homesite in the park. This tree has been planted rather extensively in some of the communities near the park where it appears to thrive.

This evergreen tree of the Coastal Plain has attractive dark glossy leaves.

4. BIGLEAF MAGNOLIA—*Magnolia macrophylla* Michx.

Jennison [1939a] stated that this tree was "Reported to occur in Le Conte Creek valley at about 2400 ft. elevation," but the identification was probably in error, and the species referred to was either *Magnolia fraseri* or *M. tripetala,* both of which occur there. The bigleaf magnolia grows within 4 miles of the park boundary near Boogertown, northeast of Gatlinburg, where it is indigenous. It has been introduced in Gatlinburg, but it is not known to occur within the park.

Flowers have been noted on the native trees near Boogertown in April. The fruits are spherical (differing from the ovoid fruits of *Magnolia fraseri* and *M. tripetala),* and the very large, eared leaves have a pronounced silvery cast to the underside.

5. UMBRELLA MAGNOLIA—*Magnolia tripetala* L.

In rich soils at low altitudes, chiefly along streams, the umbrella magnolia is a fairly common tree. Shanks [1954a] lists it as a characteristic small tree of the cove hardwoods, but ordinarily it is not common there since the highest elevation this magnolia reaches is about 3500 ft. Usually this tree grows below 2500 ft.

Flowers have been noted on the umbrella magnolia as early as April 13 (1952) and as late as May 28 (1954). The first half of May is its usual period of peak bloom. The fruits change from green to red in July; in August they are bright red in color.

This magnolia is fairly common along lower Little River, along Laurel Creek, in Cades Cove, along Abrams Creek, and elsewhere. In contrast to the Fraser magnolia, the leaves of the umbrella magnolia are pointed at both ends, and the fruits are smooth.

YELLOW-POPLAR—*Liriodendron* L.

YELLOW-POPLAR—*Liriodendron tulipifera* L.

From the lowest altitudes to approximately 4000 ft., in all except the driest situations, the yellow-poplar is one of the most abundant trees in the park. Shanks [1954a] listed it as a dominant canopy tree in cove hardwoods, in hemlock forests, and in closed oak forests. Whittaker [1956] included it in his mesic tree class. This species thins out rather rapidly above 4000 ft., but specimens occur at 4900 ft. along the old transmountain road (North Carolina side) and on Gregory Bald. At the lower altitudes many a dense stand comprised of at least 80 percent yellow-poplars marks the site of what was formerly cultivated land where corn and potatoes were grown.

Yellow-poplars are usually in flower throughout May, with an early record on April 3 (1946) and a belated flowering on June 17 (1961). These trees ordinarily bear a great quantity of seed which is harvested by squirrels and by certain birds.

Although the biggest yellow-poplar on record does not grow in the park,* this area probably has a greater number of specimens whose diameters exceed 5 ft. than are to be found anywhere. Along Buck Fork (Greenbrier area) alone there are at least six trees of that size, and the same can be said of the Ekaneetlee Creek area above Cades Cove. What appears to be the biggest sound yellow-poplar in the park, measuring 23 ft. 7 inches in circumference, stands almost on the Sevier-Cocke County line between Indian Camp Creek and Dunns Creek, north of Mt. Guyot.†

FAMILY ✦ *CALYCANTHACEAE*

CAROLINA ALLSPICE—*Calycanthus* L.

SWEETSHRUB‡—*Calycanthus floridus* L.

Except in places that are quite dry, this is a common shrub at the lower and middle altitudes. Shanks [1954a] listed it as dominant in cove hardwoods and in closed oak forests, while Whittaker [1956] grouped it with species comprising his submesic shrub union. Jennison [1939a] writes that it occurs "Along watercourses and on wooded slopes." It has been noted from the lowest altitudes to 3850 ft. (Pole Knob, above Cades Cove).

The reddish-maroon flowers may be expected from late April to early June. An early blossoming record is March 30 (1938). Some flowers have been noted throughout June and July, and there is a late flowering report on August 15 (1936).

The chocolate-colored seeds are often removed from the large brown pods by white-footed mice. These seeds are reputed to be

* The largest specimen reported to the American Forestry Association grows at Annapolis, Md. It has a circumference of 26 ft. 6 inches [Dixon, 1961].

† Buckley [1859] writes of a yellow-poplar "near the Pigeon River in Haywood County, North Carolina, about eight miles from the Tennessee line, thirty-three (33) feet in circumference at three feet from the ground, or eleven feet in diameter, and upwards of one hundred feet high. Another on the western slope of the Smoky Mountains in Tennessee, on the Little Pigeon River, is twenty-nine feet in circumference at three feet from the ground."

‡ According to Niceley, the variety *laevigatus* (Willd.) T. & G. is represented.

poisonous to livestock [Ammons, 1950; Grimm, 1957]. The leaves are rather fragrant when crushed.

FAMILY **+** *ANNONACEAE*

PAWPAW—*Asimina* Adans.

PAWPAW—*Asimina triloba* (L.) Dunal

This is an uncommon shrub or small tree that occurs rather irregularly in the park. Ordinarily it grows in colonies due to its habit of propagating by rootshoots. It is to be found in rich soils at elevations up to 2600 ft. The pawpaw grows near the Park Headquarters area, Greenbrier, Cades Cove, near the Sinks on Little River, along lower Abrams Creek, and elsewhere. A colony of these plants at 2600 ft. along the transmountain road near Cliff Branch (North Carolina side) appears to be at the highest altitude this species reaches in the park.

The brownish-maroon flowers usually appear in April; in the Park Headquarters area the fully-colored blossoms have been noted from late March to early May. Unusually late flowers were observed in Happy Valley on May 29, 1946.

Most of the pawpaws in the area do not bear fruit. I have noted ripe fruits in Fontana Village on September 3 (1953) and near the Abrams Creek Ranger Station on September 25 (1950).

What appears to be the largest pawpaw in the park, measuring 29 inches in circumference, grows approximately ½ mile southeast of the Abrams Creek Ranger Station on what is locally known as the Dock Gourley place.

FAMILY **+** *LAURACEAE*

SASSAFRAS—*Sassafras* Nees

SASSAFRAS—*Sassafras albidum* (Nutt.) Nees

This is a common tree at low and middle altitudes, especially in

old fields and disturbed areas. Shanks [1954a] lists it as a characteristic tree of open oak and pine stands, and Whittaker [1956]
includes it with his subxeric tree class. Ordinarily it is uncommon
to rare above 4000 ft., but I have noted it at 5050 ft. on Thomas
Ridge, and Whittaker's [ibid.] population chart shows it extending
to 5500 ft. in these mountains.

The first half of April marks its usual peak of flowering, but
it has come into blossom as early as March 15 (1938, 1939),
and I have noted flowers as late as May 17 (1959) at 4600 ft.
It fruits rather uncommonly. Bears appear to be quite fond of
sassafras fruits, occasionally breaking down the smaller trees in
quest of them.

Although this tree seldom attains considerable proportions in
the park, Jennison [1939a] mentions the occurrence of a few in
the Meigs Creek valley that are over 30 inches in diameter, and
Sharp informed me (1963) of specimens along Rabbit Creek that
are of approximately the same size.

WILD ALLSPICE—*Lindera* Thunb.

SPICEBUSH—*Lindera benzoin* (L.) Blume

This shrub is very common at low altitudes, especially "along
water courses or in well irrigated soil" [Jennison, 1939a]. In the
Cataloochee area it occurs up to an elevation of 2700 ft., while
along Big Creek it reaches to 2800 ft.—possibly its uppermost
range in the park.

The spicebush is one of the earliest shrubs to come into
flower, the first half of March marking its usual peak of blossoming. During the 25-year period beginning in 1938, flowers have
appeared in February in four of the years (earliest: February 13,
1950). The green fruits turn a bright glossy red color in August.
When bruised, the twigs and leaves have a strong spicy aroma.

FAMILY ✦ *SAXIFRAGACEAE*

MOCK-ORANGE*—*Philadelphus* L.

1. HAIRY MOCK-ORANGE†—*Philadelphus hirsutus* Nutt.

This uncommon shrub grows on "rocky slopes and banks of streams" [Jennison, 1939a] at low altitudes. It occurs along Little River (Sinks and below), Abrams Creek, and in Wears Cove. Flowering dates range from April 16 (1945) to May 3 (1937).

2. SCENTLESS MOCK-ORANGE—*Philadelphus inodorus* L.

According to Sharp et al. [1960], two varieties of this species occur in the park. One, *P. i. grandiflorus* (Willd.), is believed to be introduced while the other, *P. i. strigosus* Beadle, is native. There is a specimen of the former from Elkmont in the park's herbarium; this shrub was in bloom May 24, 1936. A specimen determined by Hu as *P. i. strigosus* (University of Tennessee Herbarium) was collected near Gatlinburg in 1935.

3. SHARP'S MOCK-ORANGE—*Philadelphus sharpianus* Hu

This rare shrub was named in honor of Dr. A. J. Sharp (University of Tennessee) by Dr. S. Y. Hu [1956], an authority on *Philadelphus*. It is of special significance since the specimen on which Dr. Hu based her description was found growing near the mouth of Mill Creek, on Abrams Creek, in the park.

Hu [ibid.] writes that "It grows on sandstone ledges or river

* Plants of the genus *Philadelphus* are mostly uncommon to rare within the park, even though Hu [1955], in comments relating to *P. inodorus,* says that "judging from the specimens which I have examined, the Great Smokies is the center of concentration of this species." *Philadelphus* is largely restricted to bluffs and ledges at low altitudes in the western end of the park. In addition to the species referred to, Sharp et al. [1960] indicate that *P. pubescens* Loisel. occurs in our area. I have noted that species about 3 miles north of the park boundary in Wears Cove (where it was in full flower April 16, 1945), but it does not appear to grow within the park.

† Dr. Hu, who determined the specimens of *Philadelphus* in the University Herbarium, called a specimen collected near the Sinks along Little River *P. hirsutus;* a specimen from near the mouth of Abrams Creek she named *P. hirsutus nanus.* According to Sharp et al. [1960], *P. h. intermedius* also occurs in that area, but I have found no park specimens with that determination.

bluffs as a shrub about 10 ft. high," and that "Its white flowers appear in late April or mid-May."

HYDRANGEA—*Hydrangea* L.

1. WILD HYDRANGEA—*Hydrangea arborescens* L.

This is a common shrub over much of the park. Shanks [1954a] lists it as a dominant shrub in cove hardwoods, in hemlock forests, in northern hardwoods, and in closed oak forests. Whittaker [1956] writes that it is "widespread in mesic and submesic sites." It is by far the most prevalent of the two native hydrangeas, occurring from low altitudes to 6450 ft. (Clingmans Dome).

Flowers ordinarily appear in late May or early June (earliest: May 16, 1946), with June and July being the months of usual blossoming. Occasionally a few flowers are noted in August and as late as the end of September.

2. SILVERLEAF HYDRANGEA—*Hydrangea radiata* Walt.

This shrub, readily identified by the downy silvery-white undersurface of the leaves, appears to be restricted to the extreme western end of the park where it is not uncommon in rocky situations. It has been noted in the Abrams Creek gorge, in Happy Valley, along Parsons Branch, and in the Fontana Dam area. All localities are below 2000 ft.

The flowering of the silverleaf hydrangea has been recorded in June.

SWEETSPIRE—*Itea* L.

TASSEL-WHITE—*Itea virginica* L.

This shrub is rather rare, appearing to be confined to a few localities in the western end of the park. There are no records for it east of Cades Cove where it grows in the so-called "gum swamp" near the loop road about ½ mile east of the Cable Mill. Other localities include Happy Valley, Abrams Falls, and the Abrams Creek gorge. None of these places exceeds 1800 ft. in elevation.

Flowering dates range from May 18 (1948) to June 28 (1935).

CURRANTS; GOOSEBERRIES—*Ribes* L.

1. PRICKLY GOOSEBERRY*—*Ribes cynosbati* L.

This is a fairly common shrub in rather moist soils at middle and high altitudes. Specimens have been collected from 2000 ft. (Big Creek) to near the summits of some of our highest mountains (6600 ft.). Flowers have been noted in late April and in May. When ripe the prickly fruits have a good flavor.

Large numbers of these and other species of *Ribes* have been pulled up over a long period of years in the continuing attempt to eradicate plants known to be alternate hosts for the white pine blister rust. White pines growing within 900 ft. of *Ribes* bushes become susceptible to infection with the deadly fungus.

2. SKUNK CURRANT—*Ribes glandulosum* Grauer

This is a rather uncommon shrub largely confined to cool, moist, rocky slopes at the higher elevations. Ordinarily of small size and procumbent habit, it is characterized by ill-smelling leaves and twigs. The fruits are red when ripe, hairy, and unpalatable. The skunk currant is largely a plant of the spruce-fir forest where it may reach to the summits of some of the highest mountains (6600 ft.).

The appearance of the flowers has ranged from as early as May 12 (1934) to as late as June 28 (1955).

3. MISSOURI CURRANT—*Ribes odoratum* Wendland f.

This shrub of the western United States was introduced about a few old homesites in pre-park days. It was reported eliminated in the Ravensford area by *Ribes*-eradication crews in the late 1940's. It also grew in Elkmont.

4. ROUNDLEAF GOOSEBERRY—*Ribes rotundifolium* Michx.

This is a common shrub, mostly at altitudes above 4500 ft. Shanks [1954a] lists it as one of the dominant shrubs of the spruce-fir forest. Sharp [1942b] called it "the common gooseberry in the high mountains, with small leaves and smooth fruits." The round-

---

* Jennison [1935a] published "Notes on Some Plants of Tennessee" in which he mentioned the discovery of *Ribes lacustre* (Pers.) Poir. near the summit of Mt. Le Conte. The specimens were later re-determined as *R. cynosbati*.

leaf gooseberry grows on or near the summits of some of our highest mountains (6600 ft.). The only record of this species below 4500 ft. involves one collection made at 2800 ft. near Huskey Gap.

Flowering dates range from April 27 (1945) to June 5 (1940). The fruits ripen in August.

FAMILY ✦ *HAMAMELIDACEAE*

WITCH-HAZEL—*Hamamelis* L.

WITCH-HAZEL—*Hamamelis virginiana* L.

Except in some of the driest habitats, this is a common shrub or small tree at low and middle altitudes. Shanks [1954a] lists it as one of the small trees characteristic of cove hardwood and closed oak forests. Whittaker [1956], who includes the witch-hazel in his submesic tree class, indicates that it reaches an altitude of 5900 ft., but ordinarily it is quite scarce above 5000 ft.

Flowers have been noted as early as September 21 (1953, 1960), but ordinarily it is early in October when this late-blossoming plant begins to flower. Occasionally a few flowers may be noted as late as January or even later (February 17, 1939).

FOTHERGILLA—*Fothergilla* Murr.

LARGE FOTHERGILLA—*Fothergilla major* Lodd.

This spring-blossoming relative of the witch-hazel, which it resembles in leaf but not in flower, occurs within 3 or 4 miles of the park line along Cove Creek in Wears Cove. Jennison [1939a] also reports it growing along Norton Creek, near Gatlinburg. It is a rare shrub with showy cream-colored flowers that usually appear in early or middle April.

SWEETGUM—*Liquidambar* L.

SWEETGUM—*Liquidambar styraciflua* L.

In the lower altitudes, this tree is quite common along streams and in moist situations. Locally it may be abundant. Its highest occurrence is at approximately 2000-2500 ft. The flowers of sweetgum have been noted as early as March 17 (1945), but its usual time of blossoming is April.

In October the star-shaped leaves of this tree assume a variegated array of colors. The branches of the sweetgum usually bear corky wings which in some specimens become very prominent.

FAMILY ✦ *PLATANACEAE*

SYCAMORE—*Platanus* L.

AMERICAN SYCAMORE—*Platanus occidentalis* L.

This is a common streamside tree at low altitudes and up to approximately 3200 ft. In most watersheds it drops out below the 3000 ft. elevation. Flowers appear in April and early May. The fruits are usually solitary, but Sharp found them occurring in pairs on a tree that grew along Anthony Creek in Cades Cove (December, 1951).

Specimens measuring approximately 17½ feet in circumference (at 4½ feet from the ground) grow along the West Prong of the Little Pigeon River (Chimneys Campground) and near the mouth of Buck Fork (Greenbrier area).

FAMILY ✦ *ROSACEAE*

NINEBARK—*Physocarpus* Maxim.

NINEBARK—*Physocarpus opulifolius* (L.) Maxim.

This shrub is rare in the park, there having been but a single collection from Whiteoak Sink where the plant grows "on a dolomite bluff" [Jennison, 1939a].

SPIRAEA—*Spiraea* L.

Spiraea spp.

According to Sharp [1942b], at least two species of cultivated *Spiraeas** were introduced around old homesites. Some of these shrubs may continue to grow for many years.

MOUNTAIN-ASH—*Sorbus* L.

AMERICAN MOUNTAIN-ASH—*Sorbus americana* Marsh.

Since it is only rarely that this small tree grows at altitudes below 5000 ft., the mountain-ash may be regarded as the most boreal of our deciduous trees. Ordinarily it occurs in the spruce-fir forest, especially in openings created by fire or windthrow, but it also occurs at 5000 ft. or above in localities beyond the range of the Canadian zone conifers (e.g., Spence Field Bald). Within the narrow high-altitude belt it occupies, the mountain-ash is a fairly common species. It ranges to 6600 ft. on Clingmans Dome and on Mt. Guyot.

The flower clusters which at a distance resemble the inflorescence of the American elder *(Sambucus canadensis)* have appeared as early as June 7 (1939), and some were noted as late as July 18 (1946). Normally the peak of blossoming is during the last half of June. In September the fruit clusters are a vivid orange-red or coral in color. Fruits are not produced every year; according to my records there is a tendency toward a heavy fruit crop at 3-year intervals (1943, 1946, 1949, 1952, 1955), but this regularity of recurrence sometimes goes awry.

Baldwin [Dixon, 1961] discovered one of these trees with a trunk circumference of 5 ft. 6 inches, making it the largest American mountain-ash recorded to date. It grows near the Boulevard Trail to Mt. Le Conte, approximately 3 miles east of Newfound Gap.

* *S. prunifolia; S. vanhouttei.*

The large white flowers of the mountain stewartia *(Stewartia ovata)* have numerous stamens that may be purple (above) or yellow (below) in color. This rather uncommon plant, ordinarily a shrub, appears to be restricted to altitudes between 1000 and 2500 ft.

The large blossoms of the umbrella magnolia *(Magnolia tripetala)* are white or creamy-white. The deciduous leaves are pointed at both ends.

Ordinarily the white or pinkish flowers of the mountain silverbell *(Halesia carolina monticola)* are at their peak in late April and early May. At high altitudes they may persist until early June.

(left) The well-named shagbark hickory *(Carya ovata)* is one of the rare trees in the park. This trio grows beside the old Rich Mountain road near Cades Cove.

(below, left) In pre-park days the black cherry *(Prunus serotina)* was one of the trees most sought by lumbermen. It is one of the dominant species in the cove hardwood forests.

(below, right) A mature yellow buckeye *(Aesculus octandra)* is usually characterized by its scaly bark. This is the only tree in the park with palmately-compound leaves.

This giant grapevine *(Vitis vulpina)* measured 19 inches in diameter (1935). It grew a few miles southwest of Cosby, between Old Black and Maddron Bald.

Looking westward from high up on the Alum Cave Trail to Mt. Le Conte. The vista is framed by red spruces *(Picea rubens).*

This tree is readily identified by its pinnate leaves which closely resemble those of the staghorn sumac *(Rhus typhina)*.

PEAR—*Pyrus* L.

1. RED CHOKEBERRY—*Pyrus arbutifolia* (L.) L. f.

This shrub is rare in the park, having been collected only on Brushy Mountain (4900 ft.). It has also been found approximately 1 mile north of the park line in the Glades, just west of Emerts Cove, at 1600 ft.

In October this is a very attractive plant with red fruits and with leaves that are crimson above and downy-white beneath.

2. PEAR—*Pyrus communis* L.

A few of the pear trees that were introduced in the vicinity of former dwellings may still persist. Ordinarily this cultivated species comes into bloom in March; there is one record of flowers on February 28 (1938).

3. PURPLE CHOKEBERRY—*Pyrus floribunda* Lindl.

Our only record of this plant is based on the collection of a specimen in the Oliver meadow in Cades Cove by Sharp (May 19, 1957).

4. BLACK CHOKEBERRY—*Pyrus melanocarpa* (Michx.) Willd.

Unlike the red and purple chokeberries, the black-fruited plant is not uncommon. Shanks [1954a] lists it as one of the tall shrubs in open oak and pine stands, while Whittaker [1956] includes it as a member of the lower-elevation heath bald shrub union. In the wet Oliver meadow in Cades Cove (1800 ft.) the flowering of this shrub has ranged from April 16 (1945) to June 24 (1935); on Mt. Le Conte the blossoms have been noted as late as July 17 (1940).

APPLE—*Malus* Mill.

1. SOUTHERN CRAB APPLE—*Malus angustifolia* (Ait.) Michx.

This small thorny tree is quite scarce in the park where it is confined mostly to the lower altitudes. Except for one record at

2800 ft. near the mouth of Kephart Prong, all the other localities where it has been found are at 1800 ft. or lower—Cades Cove, Happy Valley, Gatlinburg, Deep Creek, and near Fontana Dam. The large, pink, fragrant flowers have been noted from April 13 (1945) to May 4 (1937). Sharp [1942a] writes that this little tree grows in moist soils and that its sour-tasting fruits are usually less than an inch in diameter.

2. APPLE—*Malus pumila* Mill.

This foreign tree was introduced into many localities, even on some of the high-altitude grass balds where herders tended livestock in pre-park days. Jennison [1939a] said it was "widely naturalized through escape from cultivation, sometimes occurring mingled with forest trees. More often around old dwellings, in old fields and on waysides." It is gradually disappearing.*

In Cataloochee an apple tree 30 inches in diameter grew on the John Jackson Hannah farm at an altitude of 3150 ft.; by 1948 this tree, then but a shell, had but a few living branches. Apple trees grew at Mt. Sterling Gap (3894 ft.), Pin Oak Gap (4428 ft.), on Russell Field (4250 ft.), and on Spence Field Bald (approximately 5000 ft.)—the last being the highest altitude for which there is a park record.

Late March and early April is the usual time for the appearance of the flowers; earliest was on March 19 (1945). Latest blossoms were recorded on April 25 (1936).

SERVICEBERRY—*Amelanchier* Med.

1. DOWNY SERVICEBERRY—*Amelanchier arborea* (Michx. f.) Fern.

This serviceberry is closely related to the more common Allegheny serviceberry *(A. laevis)* [Jones, 1946] and could readily be confused with it.† According to Jennison [1939a], it is "known only

---

* In the 1920's M. M. Whittle began setting out an apple orchard along Le Conte Creek 3 miles southeast of Gatlinburg. "Cherokee Orchard," as it was called, eventually consisted of 7,000 apple trees of various varieties, most of which were brought in from a Knoxville, Tenn., nursery. After the orchard became a part of the Great Smoky Mountains National Park it continued to be operated on a lease arrangement for a number of years. Finally, in about 1957, it was abandoned.

† Authorities agree that it is difficult to delimit the species of *Amelanchier*. Gleason [1952], in the key to the species in this genus, states that at the time of flowering *A. arborea* has leaves that are much less than half-grown and tomentose beneath, whereas *A. laevis* has leaves that are about half-grown and smooth beneath.

in neutral soils over limestone on borderlands of park," based on the records from Dry Valley, Townsend, Rich Mountain, and Wears Cove. R. C. Burns and I collected what we believed to be this species in Happy Valley, near the park boundary, where it was in partial flower at the exceptionally early date of February 17 (1950). According to Whittaker [1956], however, this species of *Amelanchier* ranges up to 3700 ft. and represents one of the entities in his submesic tree class. In Jones' monograph [1946] reference is made to a collection of the downy serviceberry that was made on Spence Field, on the state line, near 5000 ft.

Further investigation is necessary before the status of this little tree in the park can be determined.

2. ALLEGHENY SERVICEBERRY; "SARVIS"—
   Amelanchier laevis Wieg.

This tree, the common serviceberry of the mountains, ranges from the lowlands to over 6000 ft. Gilbert [1954] calls it "the most common deciduous tree on the [grass] balds." According to Shanks [1954a], it is a dominant small tree in cove hardwoods and northern hardwoods and a typical species in spruce-fir and in open oak and pine stands. Whittaker [1956] groups it with yellow birch, mountain maple, and alternate-leaf dogwood in the "ecotonal-mesic union, centered in mesic sites at elevations around 4500 ft." For persons interested in this handsome tree it is recommended that a hike be made over the Appalachian Trail between Clingmans Dome and Silers Bald on or about May 10-15 when the Allegheny serviceberries, which are quite plentiful there and which grow to considerable proportions in that area, are at their peak of flowering.

A small specimen I noted just below Cliff Top on Mt. Le Conte at 6450 ft. was in flower May 20, 1941. This appears to mark the highest altitude reached by this serviceberry. On Clingmans Dome, at 6300 ft., I made note of a small tree in blossom on May 29, 1939; at the same altitude, along the Appalachian Trail between Clingmans Dome and Double Springs Gap, a tree of this species was coming into flower May 12, 1949.

Following an exceptionally mild winter, such as in 1950,* this

---

* Up to middle February 1950 the weather was unseasonably mild, but the month of March proved to be colder than January.

tree may come into flower before the end of February (earliest: February 21, 1950), but ordinarily it is the second or third week of March before flowers appear. In the wake of very severe winters (1940, 1941, 1942, 1947, 1958) this serviceberry may not begin to blossom before April. Flowers have been noted as late as June 5 (1940) at 6000 ft. on Mt. Le Conte. One specimen I had under observation along lower Little River for a period of 15 years (beginning in 1938) came into full flower as early as March 7 (1949, 1950) and as late as April 10 (1941). During the interval from 1939 to 1956 there were 7 years when the flowering of the Allegheny serviceberry within the park extended over a span of 10 to 12½ weeks—the longest being from February 21 to May 20 (1950).

Below 2000 ft. the fruits are ordinarily ripe in middle June, while at the upper limits of its range it may be late August before the fruits ripen.

An Allegheny serviceberry of record size grows about 60 ft. south of the Appalachian Trail at a place approximately 1 mile west of Silers Bald, at 5150 ft.; in 1949 it measured 6 ft. 2 inches in circumference and stood 60-70 ft. high. This tree was in full flower on May 12, 1949.

3. ROUNDLEAF SERVICEBERRY—*Amelanchier sanguinea* (Pursh) DC.

Although no specimens of this shrub have been noted in the park, it occurs at a low elevation on bluffs in Wears Cove within 3 miles of the boundary. It also occurs in western North Carolina at Asheville and Chimney Rock [Jones, 1946]. This broad-leafed serviceberry was in flower in Wears Cove on April 2 (1938) and May 8 (1937).

Chaenomeles Lindl.

JAPANESE QUINCE—*Chaenomeles lagenaria* (Loisel.) Koidz.

This cultivated shrub was introduced about old homesites [Sharp, 1942a]. It is now rare in the park.

HAWTHORN—*Crataegus* L.

According to Little [1953], "*Crataegus,* with perhaps between 100 and 200 species of small trees and shrubs in the United

States (nearly all in the eastern half), remains the largest and taxonomically most difficult genus of native trees. More than 1,100 specific names have been published for the native plants of this genus."

The advice given by Ammons [1950] is pertinent: "The student who desires to attempt a classification of the hawthorns should be sure to have specimens from the same plant in flower and with mature fruit. The color of the anthers and the color of the fruit should also be noted."

The determinations on which the present account is based were verified or made by Ernest J. Palmer, "the leading authority on the genus" [Little, 1953].

1. BILTMORE HAWTHORN—*Crataegus biltmoreana* Beadle

This shrub is of rare occurrence. A fruiting specimen was collected in Whiteoak Sink (approximately 1750 ft.) on August 22, 1934.

2. BOYNTON HAWTHORN—*Crataegus boyntonii* Beadle

According to Jennison [1939a], this shrub grows along lower Abrams Creek "in well drained mineral soils." It has not been recorded from elsewhere in the park.

3. PEAR HAWTHORN*—*Crataegus calpodendron* (Ehrh.) Med.

Like the preceding species, this shrub is rare in the area. The only record is from Cades Cove where the plant grew over limestone.

4. *Crataegus cibaria* Beadle†

This shrub has been recorded only from Cades Cove.

5. SANDHILL HAWTHORN—*Crataegus collina* Chapm.

Jennison [1939a] reports this shrub or small tree from sandy soil near Rich Mountain Gap. It is not known from elsewhere in the park.

---

* According to E. J. Palmer, the variety *microcarpa* (Chapm.) Palmer is represented.

† This was labeled *C. iracunda* Beadle by Jennison who collected it. In 1940 E. J. Palmer called this specimen *C. cibilis,* but in 1953 he redetermined it (#3014) as *C. cibaria.* This species does not appear in Little [1953] nor in Fernald [1950].

6. COCKSPUR HAWTHORN*—*Crataegus crus-galli* L.

Like most hawthorns, this one is rare having been found only in Cades Cove. According to Sharp [1942a], it grows there in sandy loam.

7. GATTINGER HAWTHORN—*Crataegus gattingeri* Ashe

A specimen collected on Gregory Bald by S. A. Cain on August 4, 1929, was given this name by E. J. Palmer (1953).

8. THICKET HAWTHORN—*Crataegus intricata* Lange

In Wears Cove, within 3 miles of the park boundary, I collected a flowering specimen of hawthorn on April 15, 1945, which E. J. Palmer called *C. intricata*. The elevation there is approximately 1250 ft. This species has not been found growing within the park.

9. LARGE-SEED HAWTHORN†—*Crataegus macrosperma* Ashe

This is the only one of 10 species of hawthorns that is not regarded as rare and localized in the park and vicinity. As plants go it is not common, but as hawthorns go it is by far the most prevalent. In a letter to A. J. Sharp dated August 13, 1940, E. J. Palmer wrote that "It is certainly the common form found at higher altitudes in your region and widely throughout the eastern and middle states as far south as the Piedmont." Gilbert [1954] regarded it as a common plant on Gregory Bald where both the typical form and the variety *roanensis* have been collected. The latter variety also grows on Spence Field Bald, Parson Bald, Silers Bald, Andrews Bald, Heintooga Bald, High Rocks (summit), Raven Fork (at 3000 ft.), Thunderhead (summit), the Sinks on Little River (1565 ft.), and in Big Cove near Cherokee, North Carolina. In addition to Gregory Bald the typical variety has been collected at Halls Cabin above Tremont, on Steel Trap Branch, and at Deals Gap. At the last-named place (approximately 1900 ft.) the plant was in flower on May 1, 1936.

---

* According to E. J. Palmer, both the typical form and the variety *macra* (Beadle) Palmer are represented in the park.

† According to E. J. Palmer (1953), both the typical form and the variety *roanensis* (Ashe) Palmer occur here, the latter more likely to be found at high altitudes.

10. PIEDMONT HAWTHORN—*Crataegus regalis* Beadle

Like most hawthorns this one is of rare occurrence here. In August 1934 a specimen was collected on the state line east of Gregory Bald.

BRAMBLE—*Rubus* L.

While examining the specimens of *Rubus* in the University of Tennessee Herbarium I was impressed by the frequency with which L. H. Bailey, who had made a study of this baffling genus for over half a century, used the words "perhaps" or "probably" before the names he affixed to each one. (The specimens had been sent to him for naming.) This, as most authorities agree, is a particularly difficult group of plants. The reasons for this and the problems involved are discussed by Gleason [1952: Vol. 2, p. 305]. Sharp et al. [1960] listed 30 species for the state of Tennessee. The 15 species (more or less) listed for Great Smoky Mountains National Park in the present account are grouped into raspberries, 3; dewberries, 5; and blackberries, 7.

Of the raspberries, whose fruit when picked has a hollow center, the three species are readily identified. (1) The red raspberry *(Rubus idaeus* var. *canadensis)* appears to be restricted to that part of Clingmans Dome which is above the 6000 ft. altitude. It has white flowers, delicious red fruits, and canes that are covered with as many soft bristles as there are hairs on a dog's back. (2) The black raspberry *(R. occidentalis),* restricted to altitudes below 2000 ft., is "the only raspberry in the park with purplish-black fruit" [Sharp, 1942b]. (3) The purple-flowering raspberry *(R. odoratus)* is our only *Rubus* with simple instead of compound leaves. These leaves are large, three- to five-lobed, and they resemble the leaves of the striped maple *(Acer pensylvanicum).* The attractive rose-purple flowers somewhat resemble those of a wild rose. Ordinarily the red fruits are rather insipid. The leaves of (1) and (2) are white or grayish beneath.

Five species of dewberries are included. These are plants whose first-year canes are prostrate or low-arching and which normally take root at the tip.

The canes of blackberries are usually erect. Of the seven species listed, three were described by L. H. Bailey who regularly reviewed the earlier collections in the University of Tennessee and "who erected several species on the basis of Tennessee material; subsequent treatment has been more conservative" [Sharp et al., 1960].

1. RED RASPBERRY—*Rubus idaeus* L. var. *canadensis* Richards.

As indicated above, this bristly-stemmed plant appears to be restricted to Clingmans Dome where it occurs commonly among the rocks at altitudes above 6000 ft. At that high elevation the environment consists of frequent fog and heavy precipitation (averaging 85 inches per year), low winter temperatures, and a short growing season.

The white flowers appear in June and July, and the excellent red fruits begin to ripen in late July and early August. The latter part of August and on to middle September finds the fruits fully mature.

"This species resembles the commercial red raspberry but the stems are much more bristly" [Sharp, 1942b].

2. BLACK RASPBERRY*—*Rubus occidentalis* L.

This is an uncommon plant with records of occurrence at Greenbrier, in the Sugarlands, and along Twentymile Creek. It has not been noted above 2000 ft. The fruit was ripe on May 28 (1937). As Sharp [1942b] stated, no other raspberry in the park has purplish-black fruit. The undersurface of the leaves is whitish.

3. PURPLE-FLOWERING RASPBERRY—*Rubus odoratus* L.

This attractive and unique species of *Rubus* with simple maple-like leaves, large rose-purple flowers, and spineless branches is a common shrub at low and middle altitudes where it occurs in rather rich soils. It grows to 5000 ft. elevation in the Flat Creek area. Flowers may appear by middle May in the lower altitudes, and some blossoms have been noted throughout the summer months and on as late as October 13 (1938). Therefore it is not unusual

* According to Shanks [1961], the form *pallidus* (Bailey) Robins. is included with one collection from Greenbrier Cove.

to find flowers and ripe fruits at the same time. The fruits of this raspberry are rather insipid, but they are quite edible.

4. BAILEY DEWBERRY—*Rubus baileyanus* Britt.

Sharp [1942b] wrote that this "or a closely related species has been found at Elkmont and in the Greenbrier Cove."* Both Sharp et al. [1960] and Shanks [1961] listed this dewberry as a park entity.

5. ENSLEN DEWBERRY—*Rubus enslenii* Tratt.

This appears to be a rare dewberry as it has been recorded only from Rich Mountain and from along the trail to Mt. Cammerer. Jennison [1939a], Sharp [1942b], and Shanks [1961] included this species in their lists of park plants. Flowering specimens were collected on April 30 (1935) and May 20 (1928).

6. NORTHERN DEWBERRY†—*Rubus flagellaris* Willd.

Jennison collected this plant near Bryson City, North Carolina, and Sharp [1942b] discovered it in the Abrams Creek gorge. It grows "in sterile sandy soil at low elevations" [Jennison, 1939a]. There are no other records. A specimen collected May 3, 1937, is in flower.

7. SWAMP DEWBERRY—*Rubus hispidus* L.

This plant was collected in Whiteoak Sink and in Cades Cove "in wet acid sandy soil" [Jennison, 1939a]. It has not been recorded from elsewhere in the park. Sharp [1942b] described it as "a prostrate, half-evergreen shrub forming a ground cover."

8. SOUTHERN DEWBERRY—*Rubus trivialis* Michx.

This dewberry appears in lists of park plants by Jennison [1939a], Sharp et al. [1960], and Shanks [1961]. It is recorded from Cades Cove and in "thickets, open fields and woods at lower elevations" [Jennison, 1939a].

---

* Deam [1940] who in 1917 reported *Rubus baileyanus* from several Indiana counties later referred it to *R. flagellaris*.

† See preceding footnote.

9. ALLEGHENY BLACKBERRY—*Rubus alleghaniensis* Porter

This blackberry has been recorded from a number of places including Whiteoak Sink, Elkmont, Cove Mountain, Gatlinburg, and Greenbrier. Jennison [1939a] calls it a plant of "roadsides and open woodlands; rocky or sandy loam." Flowers have been noted on May 24 (1934) and fruits on July 16 and August 5 (1935).

10. HIGHBUSH BLACKBERRY*—*Rubus argutus* Link

Collections of this plant were made in Greenbrier and Cataloochee, the highest from an altitude of 3500 ft. Flowers were recorded on May 15 (1934) and June 4 (1937).

11. THORNLESS BLACKBERRY†—*Rubus canadensis* L.

This is an abundant plant at high altitudes. Shanks [1961] lists it as one of the typical shrubs in hemlock forests, in northern hardwoods, and in spruce-fir forests. Oosting and Billings [1951] found this species present in every one of the nine stands of spruce-fir forest they studied in the park, all of which are above 5450 ft. It grows on all the grass balds and probably in every sizable forest opening at the higher elevations. Sharp [1942b] states that it occurs in moist soils above 3000 ft.; the summits of the highest mountains determine its uppermost limits.

Flowers appear throughout June and July; this species was in blossom near the summit of Mt. Le Conte as late as July 31 (1929). Late-flowering plants will not ripen their fruits before the advent of the early frosts at these high altitudes.

Although essentially thornless, this blackberry often develops small spines along the stems. A cane believed to be of this species of blackberry measured 27 ft. in length.

* The variety *scissus* Bailey is included in accordance with determinations made by L. H. Bailey.

† Bailey [1945] refers to a specimen of blackberry from Elkmont which he determined as *R. immanis* Ashe. He states that it looks like *R. canadensis* and may be related to it: "We are likely to revise the definition of this species [*immanis*] when its territory is well collected." In view of Bailey's remarks, my failure to locate an herbarium specimen, and since no mention is made of that species in the standard manuals, I am omitting *R. immanis* from the present list.

12. JENNISON BLACKBERRY*—*Rubus jennisonii* Bailey

Bailey [1945] who dedicated this plant to the memory of the late H. M. Jennison includes a drawing of it on page 637 of his description of the new species. Specimens are from Wears Cove, Gatlinburg, Fighting Creek valley, and Meadow Branch (Blount County, Tennessee). Habitats given are moist situations on sunny open slopes and along roadsides.

13. SMALL BLACKBERRY—*Rubus pauxillus* Bailey

This species is included on the basis of a single collection made in the Park Headquarters area on May 24, 1937, when the plant was in flower. Dr. Bailey made the determination.

14. TENNESSEE BLACKBERRY—*Rubus tennesseanus* Bailey

This was described as a new species of blackberry by Bailey [1934b] who collected the type specimen in 1933 on Rich Mountain at 2000 ft. altitude. Bailey writes that this new species resembles the thornless blackberry *(R. canadensis)*. Illustrations of *R. tennesseanus* appear in Bailey's 1934 (p. 270) and 1944 (p. 553) accounts of this plant. Additional locations for this blackberry are Big Creek and Whiteoak Sink.

15. TRUCULENT BLACKBERRY—*Rubus trux* Ashe

An illustration of this plant is to be found in one of Bailey's [1945, p. 681] reports along with a statement that he collected it in Cades Cove. He describes it as an "Imperfectly understood species in mountains of western North Carolina and eastern Tennessee."

Kerria DC.

KERRIA—*Kerria japonica* (L.) DC.

This cultivated shrub persists in the Greenbrier area and possibly in a few other localities where it had been introduced about former homesites.

* Gleason [1952] groups *Rubus jennisonii* with the collective species of *R. argutus*. Bailey published the description of *R. jennisonii* after the publication of the latest edition of Small [1933], and it is south of the range of plants covered by Fernald [1950].

ROSE—*Rosa* L.

1. ARKANSAS ROSE—*Rosa arkansana* Porter

A specimen to which this name was applied (W. H. Lewis, 1957), collected at an altitude of 3500 ft. in the Big Creek area, is in the University of Tennessee Herbarium. According to Fernald [1950], this rose occurs west of the Mississippi River.

2. CAROLINA ROSE—*Rosa carolina* L.

A specimen of this rose from along the old Cooper road in the Cades Cove vicinity was collected in flower on June 16, 1937. The habitat is given as "in damp soil near creek" and the elevation, 1800 ft. Like all wild roses, this one is rare in the park.

3. SWEETBRIER ROSE—*Rosa eglanteria* L.

This cultivated rose was found in flower on July 13 (1935) at 4500 ft. close to the park line near Black Camp Gap [Sharp, 1942b]. Jennison [1939a] stated that this exotic was formerly planted in gardens and established itself in the vicinity of dwellings.

4. SWAMP ROSE—*Rosa palustris* Marsh.

Sharp [1942b] writes that this wild rose usually grows below 3000 ft. in very moist soils. Specimens have been collected at Mt. Sterling, along the Oconaluftee River, along the transmountain road (1 mile from the loop tunnel), and at Woody's Branch in Swain County, North Carolina. Some of these roses were in flower June 27 (1934) and July 30 (1937).

5. WOOLY PRAIRIE ROSE—*Rosa setigera* var. *tomentosa* T. & G.

The only record is based on a collection made near Bull Cave on Rich Mountain in 1934. Sharp et al. [1960] and Shanks [1961] include the prairie rose *(Rosa setigera* Michx.) as one of the park plants, but there appears to be no specimen known from closer than Sevierville—about 10 miles north of the park.

PLUM; CHERRY—*Prunus* L.

1. AMERICAN PLUM—*Prunus americana* Marsh.

Along stream banks, fencerows, and borders of woodlands this small tree is of fairly common occurrence. Its altitudinal range extends up to approximately 3750 ft. (above Cataloochee Creek). As a result of sprouting from the roots, this plum ordinarily forms dense thickets.

During a mild winter flowers have appeared as early as January 24 (1937), and there are two records for February, but ordinarily it is in March that this little tree comes into blossom. The fruits ripen in July.

From the Chickasaw plum *(P. angustifolia)*, the only other wild plum in the area, the American plum can be distinguished by its much larger flowers and leaves.

2. CHICKASAW PLUM—*Prunus angustifolia* Marsh.

According to Sharp [1942a], this small tree grows in moist sandy or rocky soils near the park boundary. Its white flower petals are approximately one-half the size of the petals of the American plum, and whereas the leaves of the American plum are 3 to 4 inches long and half as wide, those of the Chickasaw plum are 1 to 2 inches long and ⅓ inch to ⅔ inch wide. Jennison [1939a] and Shanks [1961] both list the Chickasaw plum from the park, but specific data is lacking and further investigation is required before the true status of this little tree can be given.

3. MAZZARD—*Prunus avium* (L.) L.

(See *P. cerasus*.)

4. SOUR CHERRY—*Prunus cerasus* L.

Both these exotic trees are known to have occurred in the Cataloochee area, and one or both were introduced into Cades Cove, Greenbrier, Elkmont, and possibly elsewhere. Some specimens may persist to the present day.

5. HORTULAN PLUM—*Prunus hortulana* Bailey

This small tree was listed by Jennison [1939a] as being widely distributed in moist rich loams at lower elevations. Shanks [1952] mentions its occurrence in Blount and Sevier Counties, Tennessee, but not in the park. In

1962 Sharp was of the opinion that if it was present here it had been introduced. Specimens determined by H. L. Sherman (1959) indicate that it occurred in Smokemont, Elkmont, and near Gatlinburg. This plum resembles the wildgoose plum *(P. munsoniana)* and may be confused with it. There are three flowering dates ranging from April 2 to 22 (1936).

6. WILDGOOSE PLUM—*Prunus munsoniana* Wight & Hedr.

In 1962 Sharp informed me that this little tree had probably been introduced, belonging west and southwest of the park. It occurred along roads and near old homesites at the lower altitudes. Herbarium specimens from the Park Headquarters and Gatlinburg areas reveal its flowering from March 8 (1937) to March 30 (1936). This plum resembles the preceding species *(P. hortulana)*, with which it may be confused.

7. PIN CHERRY; FIRE CHERRY—*Prunus pensylvanica* L. f.

At middle and high altitudes, in localities where fire or other disturbances have affected the climax forest in recent decades, the pin cherry is an abundant tree occasionally covering such areas in almost a pure stand. This situation may prevail for perhaps a quarter of a century after which this relatively short-lived cherry begins to die out at a rapid rate. It is the pioneer tree species to invade areas where the shallow-rooted spruces and firs have been toppled by strong winds. From the summits of the highest mountains it extends down to approximately 2500 ft. Shanks [1954a] lists the pin cherry as a characteristic species of hemlock forests, northern hardwoods, and spruce-fir forests.

Flowers usually appear in late April (earliest: March 31, 1945); most of the flowering is in May with latest blossoms noted on June 12 (1940). The small red fruits, eaten by birds and bears, are ripe in August.

A specimen growing at 4000 ft. altitude along the original transmountain road on the North Carolina side is 5 ft. 4 inches in circumference.* Sargent [1933] writes that this is a tree 30 to 40 ft. high, but a storm-toppled pin cherry I measured at 2700 ft. along the Cherokee Orchard-Trillium Gap Trail was 91 ft. high, which may be a record height for this species.

This is the common wild cherry of the high altitudes, characterized by a smooth reddish-brown bark containing numerous

---

* This is larger than the former record specimen (4 ft. 7 inches) that is credited to Great Smoky Mountains National Park [Dixon, 1961].

horizontally-elongated lenticels. The simple oblong-lanceolate leaves turn an attractive pinkish-red in the autumn—in contrast to the black cherry *(P. serotina)* whose leaves turn a light silvery-green.

8. PEACH—*Prunus persica* Batsch

This widely cultivated fruit tree was introduced into many areas in the park where some specimens still persist. Since peach trees usually come into flower in March, they are subject to killing frosts in most years, and the chance of a fair to good crop of fruit is quite poor. The highest altitude at which this species continues to grow is 3650 ft., at Wild Cherry Branch in the valley of the Oconaluftee River. It is gradually disappearing from within the park.

9. BLACK CHERRY—*Prunus serotina* Ehrh.

This large forest tree is fairly common in the park occurring from the lowest altitudes to 5300 ft. (Heintooga Overlook). Shanks [1954a] listed it as dominant in hemlock forests, cove hardwoods, and northern hardwoods, and occasional in spruce-fir forests. Whittaker [1956] includes the black cherry in his mesic tree class. According to Ayres and Ashe [1905], this species comprised 5 percent of the forest trees in the Big Creek basin.

Flowers usually appear in late April or during May, the earliest date being April 3 (1946) at a low altitude; the latest date of flowering is June 17 (1961) when this cherry was going out of bloom at 4500 ft. Ripe fruit was noted littering the ground on September 17 (1943) at 4900 ft. on Thomas Ridge; on September 18 (1945) at 4200 ft. near Arch Rock; and on September 26 (1941) at 3800 ft. along the Ramsey Cascades Trail.

There are a number of large specimens widely scattered throughout the area. Along the old Indian Gap road (Tennessee side) at 4400 ft. stands a burly tree 13 ft. in circumference, and Jennison [1938] mentions two "in the woods above Sugarlands Valley which measure 4 to 5 feet in diameter." There are a number of tall mature trees in the "Cherry Orchard" (along the trail to Ramsey Cascades), the largest of which measures 11½ ft. in circumference. Along Surry Fork, a tributary of Roaring Fork flowing northwest from Trillium Gap, is a black cherry stump measuring 15 ft. 7 inches in circumference. Baldwin [1948b] has a photograph of a tree measuring 12 ft. 4 inches in circumference;

in 1950 he and Zenith Whaley located a black cherry in the Buck Fork area of Greenbrier that was 13 ft. 7 inches in circumference.

This is one of the most valuable timber trees, its mahogany-like wood being very highly prized for furniture and other uses. In pre-park days it was one of the most sought-after species, and many trees were cut by the loggers. The ruggedness of the terrain presented such serious difficulties in the projected removal of some of these trees that they were spared.

10. COMMON CHOKECHERRY—*Prunus virginiana* L.

This little tree was discovered in the park by the late R. E. Shanks on May 10, 1953, "forming thickets on rocky south-facing slope in open basswood-buckeye-sugar maple stand above Ft. Harry Cove." Two days later Dr. Shanks informed me that "this is our first definite specimen from the state" (Tennessee). The location, at an altitude of 3400 ft., is approximately ¼ mile above the Buckeye Nature Trail and about 150 yards from the trans-mountain road. At the time of its discovery this rare cherry was in late flower and early fruit. The chokecherry has not been found in any other locality in the park.

When compared with the black cherry, the chokecherry has leaves that are shorter and broader and with more teeth per inch of leaf margin. The racemes of white flowers are usually longer in the black cherry than in the chokecherry.

FAMILY ✦ *LEGUMINOSAE*

Albizzia Durazz.

SILKTREE; "MIMOSA"—*Albizzia julibrissin* Durazz.

This exotic tree is rare and localized. It grows near the boundary a few miles west of Forney Creek (Chambers Creek) and is well established and spreading within a few hundred yards of the park at the eastern end of Gatlinburg. It probably occurs close to the boundary at other locations.

KENTUCKY COFFEETREE—
Gymnocladus Lam.

KENTUCKY COFFEETREE—*Gymnocladus dioicus* (L.) K. Koch

One specimen of this introduced tree grows on the site of the old Hearon place along the Cooper road, approximately 1 mile from the Abrams Creek Campground.* Here the species is outside its natural range which extends eastward to Middle Tennessee [Munns, 1938].

HONEYLOCUST—*Gleditsia* L.

HONEYLOCUST—*Gleditsia triacanthos* L.

This tree is quite scarce in the park, the Appalachian Mountains appearing to determine the eastern limits of its natural range [Munns, 1938]. On the Tennessee side it grows along Abrams Creek and Laurel Creek and in Cades Cove and Gatlinburg; on the North Carolina side it occurs at Ravensford and along Forney Creek. All these places are below 2000 ft. in elevation. A honeylocust was recorded flowering in Ravensford on May 20 (1943).

This is the only tree in the park with large branched thorns and with brown fruits in the form of a bean pod a foot or more in length.

REDBUD—*Cercis* L.

EASTERN REDBUD—*Cercis canadensis* L.

Except in Cades Cove and along lower Abrams Creek the eastern redbud is a rather uncommon tree, but in limestone areas close to the park (Tuckaleechee Cove, Wears Cove, and Rich Mountain) it is abundant. It is a plant of the lower elevations, Whittaker [1956] giving 2200 ft. as its highest occurrence.

The attractive magenta-colored flowers usually appear during the latter half of March or in early April, the earliest being February 17 (1950); the latest were noted at the end of April 1962. So plentiful does the redbud grow on the north side of Rich Mountain (sloping toward Tuckaleechee Cove) that a drive over

---

* In July 1962 this tree was approximately 8 inches in diameter and 30 ft. high; it grows close to a butternut tree *(Juglans cinerea)*.

the tortuous old road from Tuckaleechee Cove to the park boundary at the time when this little tree is in full flower is one of the floral spectacles of the year.

YELLOWWOOD—*Cladrastis* Raf.

YELLOWWOOD—*Cladrastis lutea* (Michx. f.) K. Koch

There are a number of localities in the park where this tree grows, but for the most part it is uncommon and localized. In altitudinal range it occurs from 1700 ft. to approximately 3550 ft. The largest number of these beautiful trees occurs in the Sugarlands Valley from 2000 to 3000 ft., with several to be encountered along the course of the Big Locust Nature Trail (Chimneys Campground area) and along the Huskey Gap Trail (two fine stands each with a dozen or more trees at 2700-2900 ft.). Yellowwoods also grow near Cosby Creek (Cosby Nature Trail), along Baxter Creek in the Mt. Sterling area, along Twentymile Creek, near the Bull Head Trail (3550 ft.), on Rich Mountain, in Cades Cove, and in Greenbrier.

This is a tree of rich rocky coves. Shanks [1954a] has it listed as a non-dominant species of the cove hardwoods, and Whittaker [1956] includes it in his mesic tree class.

The flowering period usually extends throughout May, with extreme dates of April 16 (1945) and June 5 (1937). Yellowwoods are rather erratic in their blossoming with some years producing little or no bloom while a heavy spectacular flowering has been recorded at irregular intervals of 1 to 5 years (heavy in 1941, 1945, 1947, 1952, 1954, 1958, 1962, 1963). The long white panicles bear a close resemblance to the handsome racemes of wisteria. These give way to clusters of fruits that resemble the fruits of the ash. The leaves are pinnately compound, the leaflets being alternate and much larger than the opposite-arranged leaflets of the black locust. The bark of the yellowwood is black and smooth. "The wood when first exposed is bright yellow, later turning brown" [Sharp, 1942a].

BROOM—*Cytisus* L.

SCOTCH BROOM—*Cytisus scoparius* (L.) Link

A specimen of this foreign shrub has been growing for many years on the steep bank below the overlook at Newfound Gap (5040 ft.) where it appears to thrive. The Scotch broom is not known to occur elsewhere in the park.

Amorpha L.

1. FALSE INDIGO*—*Amorpha fruticosa* L.

This appears to be a rather rare shrub, having been recorded at low altitudes in moist sandy soils in Cades Cove, Happy Valley, at Abrams Falls, and in the Abrams Creek gorge. In Cades Cove it grows in the wet Oliver meadow along with the buttonbush *(Cephalanthus)* where it was flowering May 20-28 (1938), June 16 (1940), and July 19 (1942).

The foliage of this plant resembles that of the black locust and the purplish-blue flowers are arranged in spike-like racemes.

2. MOUNTAIN INDIGO—*Amorpha glabra* Desf.

This rare shrub has been noted on bluffs along the Tuckasegee River (near the mouth of Noland Creek) and in Cataloochee. Dates of flowering are May 16 (1935) and June 4 (1937). Sharp [1942b] stated that this species, in contrast to *A. fruticosa,* is essentially smooth with very short calyx lobes.

LOCUST—*Robinia* L.

1. BOYNTON LOCUST—*Robinia boyntonii* Ashe

Like other low-growing pink-flowered locusts, this Southern Appalachian endemic is quite uncommon in the park. It has been recorded from Elkmont, Cove Mountain, and from 3200 ft. at the "base of Rocky Spur, above Gatlinburg." The last-named locality

---

* According to Sharp [1942b], the variety *tennesseensis* (Shuttlew.) Palmer also occurs here having been "collected in the same areas as the species." There are specimens in the herbaria from lower Abrams Creek and Happy Valley, at low altitudes.

represents its upper limit in these mountains. It has also been noted on Chilhowee Mountain near the park's western boundary. Flowering dates range from May 12 to 24. The Boynton locust is a smooth-stemmed species.

2. BIGFLOWER LOCUST—*Robinia grandiflora* Ashe

Like the Boynton locust, the bigflower species is a shrubby pink-flowered plant that is quite scarce in the park. It has been noted in the Gatlinburg, Park Headquarters, Ravensford, and Mt. Sterling areas where the flowering dates range from May 10 to 17. The stems of the bigflower locust are very hairy. This, like the preceding species, is a Southern Appalachian endemic.

3. BRISTLY LOCUST—*Robinia hispida* L.

Jennison [1939a] states that this is a plant of well-drained sandy soils. According to Sharp, this showy hairy-stemmed shrub was possibly introduced here. It has been recorded from Ravensford, Mt. Sterling, Little Cataloochee, and from near Gatlinburg. Flowering dates range from May 12 to 26.

4. KELSEY LOCUST—*Robinia kelseyi* Hutchins

This is another of the attractive pink-flowered shrubby locusts— a group not too well understood. Both Sharp et al. [1960] and Shanks [1961] include the Kelsey locust as a park species, but I was unable to locate specimens in the park or in the University of Tennessee Herbarium. Small [1933] gives its range as the Blue Ridge Mountains of North Carolina, while Little [1953] restricts it to the mountains of western North Carolina and eastern Tennessee. Bailey [1937] calls it a shrub with slender prickles and smooth branchlets. Flowering is in May-June.

5. BLACK LOCUST—*Robinia pseudoacacia* L.

This is a common tree at low and middle altitudes especially in second-growth forests. Shanks [1954a] lists it as a canopy tree in both the closed oak forests and in open oak and pine stands. Whittaker [1956] who includes the black locust in his subxeric tree class gives 5300 ft. as its uppermost limit. I have noted it at 4900 ft. on the Noland Divide, at 4950 ft. near the Cosby Knob Trail Shelter, and at 5200 ft. on Heintooga Bald.

May is the usual time of flowering, with extreme dates of April 5 (1945) and June 17 (1946). Like the yellowwood *(Cladrastis)*, the black locust is rather erratic in its flowering, but there is a tendency to produce a heavy bloom at intervals of 3 or 4 years. Heavy flowering in these two related trees coincided in 1941, 1945, and 1952.

The Big Locust Nature Trail (Chimneys Campground area) gets its name from the splendid large specimen of this tree along the short route of the trail. This impressive giant, with rough deeply-furrowed bark, is 4 ft. 4 inches in diameter—the biggest black locust known in the park and, unlike the majority of this species, a member of the original forest.

For a number of years prior to the record-breaking cold weather of February 1958* the black locusts in the park and vicinity were plagued by a small beetle, the larvae of which subsisted on the chlorophyll in the leaves. So abundant were these insects that mountainsides where black locusts prevailed turned brown in the early summer. Since this extensive infestation was not apparent after 1957, it is assumed that the low temperatures in February 1958 proved to be a natural control of these beetles.

WISTERIA—*Wisteria* Nutt.

WISTERIA—*Wisteria frutescens* (L.) Poir.

This high-climbing vine grows at two old homesites in the Cades Cove area —one near Anthony Creek and one along Parson Branch (the latter location is approximately ¾ mile beyond the gate on the truck trail). In all probability this attractive plant was introduced. Flowers were noted May 29 (1958).

Pueraria DC.

KUDZU-VINE—*Pueraria lobata* (Willd.) Ohwi

That this foreign vine must have been a very popular plant in pre-park days is borne out by the fact that through August 1962, a total of 71 plots representing an aggregate of 66 acres had been located and sprayed in a

* At Park Headquarters (elevation 1460 ft.) the minimum temperature on the mornings of February 17, 18, and 19 was —11°, —13°, and —9°, respectively.

continuing attempt to eradicate it from the park. Large concentrations of
this fast-growing plant persist north of Fontana Reservoir, especially along
Hazel Creek. Sharp informed me (1963) that he believed most of the
kudzu-vine was introduced here about 1930-35.

FAMILY ✦ *RUTACEAE*

PRICKLY-ASH—*Zanthoxylum* L.

COMMON PRICKLY-ASH—*Zanthoxylum americanum* Mill.

This shrub or small tree is out-of-range in the park. It was introduced,
probably for its medicinal properties, at the Walker Sisters' place in Little
Greenbrier and at two localities in Cades Cove—the George Powell place
along Parson Branch and the Wilson place (between the Abrams Falls
Parking Area and the Elijah Oliver homesite).

FAMILY ✦ *SIMAROUBACEAE*

AILANTHUS—*Ailanthus* Desf.

AILANTHUS—*Ailanthus altissima* (Mill.) Swingle

This exotic tree was introduced in the Big Creek area in pre-park days;
in the vicinity of the campground is a stand of approximately ½ acre of
these trees resulting from their spread by means of basal suckers. On the
opposite (western) extremity of the park the ailanthus grows near the
boundary at the mouth of Abrams Creek and 2 miles south of there near
the confluence of Shop Creek with the Little Tennessee River. The only
other locality from which it has been reported is at approximately 2200 ft.
altitude along Lynn Camp Prong, above Tremont.

More than 75 years ago, in his remarks relating to the status of the
ailanthus in Tennessee, Gattinger [1887] stated: "Perfectly naturalized.
Widely spreading over the State."

FAMILY ✦ *BUXACEAE*

BOX—*Buxus* L.

COMMON BOX—*Buxus sempervirens* L.

This evergreen exotic was introduced in a number of areas where specimens still persist. In Cherokee Orchard it was cultivated at the nursery maintained for many years by M. M. Whittle.

FAMILY ✦ *ANACARDIACEAE*

SUMAC; POISON IVY—*Rhus* L.

1. FRAGRANT SUMAC—*Rhus aromatica* Ait.

Although it has not been reported from within the park, this shrub grows within ½ mile of the boundary near Fontana Dam—at the edge of a clearing in the area of the scenic overlook, at approximately 1700 ft. altitude.

2. SHINING SUMAC; WINGED SUMAC—*Rhus copallina* L.

This is an abundant and widely distributed shrub especially in clearings at low altitudes. Jennison [1939a] calls it a plant of "thickets and open woods in well drained often rocky soil." Gilbert [1954] records it from 4700 ft. on Parson Bald, but it is of rather uncommon occurrence at that altitude. The period of peak flowering is usually in late July; flowers have been noted from mid-July to mid-August.

Evening grosbeaks and several members of the thrush family have been observed feeding on the fruits.

Specimens occasionally become large enough to be called trees. Jennison mentioned one at the Dan Myers place in Cades Cove with a trunk diameter of 6 inches.*

The shining sumac is readily distinguished from other sumacs in our area by the winged stems and the drooping fruit panicles. Like the other sumacs, its leaves become bright red in the late summer and in autumn.

* From data on herbarium label #2912, University of Tennessee. (Plant collected September 12, 1936.)

3. SMOOTH SUMAC—*Rhus glabra* L.

This is a fairly common shrub "in well drained soils bordering woods and in thickets" [Jennison, 1939a]. It ranges from the lowest altitudes to 4000 ft. Flowering takes place in early summer.

From the similar-appearing staghorn sumac the smooth sumac can be distinguished by its smooth branches and very short hairs on the fruits. In autumn the showy red fruits are more brightly colored in the smooth sumac than in the staghorn.

4. STAGHORN SUMAC—*Rhus typhina* L.

This sumac is a fairly common and widely distributed shrub at low and middle altitudes. Like the other sumacs it sends up shoots from its roots forming clumps or thickets. The staghorn sumac grows on open hillsides, along roads, and in old fields. It occurs up to at least 5100 ft. altitude (between Newfound and Indian Gaps). Flowering takes place in June and July.

The staghorn sumac is more likely to reach arborescent proportions than other species of sumac. In Wild Cherry Hollow, at 3600 ft. in the upper part of the Oconaluftee River watershed, is a specimen 2 ft. 11 inches in circumference; it is 25 ft. high and has a spread of approximately 18 ft.*

From the rather similar-appearing smooth sumac the staghorn sumac can be readily identified by its very hairy branches and relatively long hairs on the fruits.

The poison sumac *(Rhus vernix)*† is of rare and local occurrence in northeastern Tennessee [Shanks, 1952]; it does not grow in the park or vicinity.

5. POISON IVY—*Rhus radicans* L.

This vine is of common occurrence at the lower altitudes where it grows in various situations but especially along fencerows, roadsides, and streamcourses. It is scarce or absent above 3000 ft. Flowers have been noted from May 14 to June 2. The small grayish-white fruits are eaten by some birds (myrtle warbler, pileated woodpecker). A freshly-killed wood-rat *(Neotoma)* I picked up near the Sinks on Little River on October 21, 1961,

---

* The largest recorded specimen, 3 ft. 4 inches in circumference, grows in Clawson, Mich. [Dixon, 1961].

† Called *Toxicodendron vernix* by Little [1953].

had been transporting a fruit-laden stem of poison ivy when it was struck by the wheels of some vehicle.

Almost invariably the leaflets occur in threes, but the size, shade of green, shape, and degree of glossiness of the leaves may differ appreciably from place to place. The leaves assume very attractive colors in the autumn. The stems of the high-climbing plants may become shaggy due to the prevalence of aerial roots. The entire plant is poisonous and should be avoided.

The poison oak *(Rhus toxicodendron)* which resembles poison ivy and which is often confused with it does not grow in the park or vicinity.

FAMILY ✦ *AQUIFOLIACEA*

HOLLY—*Ilex* L.

1. GEORGIA HOLLY—*Ilex longipes* Chapm.*

On July 9, 1953, I discovered a specimen of deciduous holly that bore its green fruits on stalks that were ½ inch long. The locality is along the trail from the Clingmans Dome Parking Area to Andrews Bald, at an altitude of about 6100 ft. The habitat with its prevalence of large rocks resembles a talus slope. This rare holly has not been found elsewhere in the park.

2. MOUNTAIN WINTERBERRY†—*Ilex montana* T. & G.

This deciduous holly is a fairly common shrub or small tree. Whittaker [1956] lists it as a member of his submesic tree class, and Shanks [1954a] includes it as a small tree characteristic of

---

* This was determined as *Ilex collina* by F. W. Woods in February 1955. Later, however, Edwin [1957] showed that *collina* was not a good species.

† On dry hillsides at low altitudes (although rarely extending upward to 4000 ft.) where low-growing blueberries are usually the dominant shrubs, we find a small deciduous holly with dense whitish pubescence on the underside of the leaves. This shrub, listed as *Ilex beadlei* Ashe in Small [1933], is relegated to the status of a variety of *I. montana* in Gleason [1952]; Little [1953] includes it under *I. montana*. This fairly common holly is listed by Whittaker [1956] as a member of the "xeric shrub union." The latter part of May is its usual time of flowering.

cove hardwoods, northern hardwoods, hemlock, and spruce-fir forests. The mountain winterberry ranges from 1800 ft. (Cove Mountain Trail) to 6100 ft. (near Andrews Bald); it is more plentiful above 4000 ft. than below that altitude. Flowering is at its peak during the first half of June, but blossoms have been noted from May 14 (1944, at a low altitude) to July 11 (1940, on Mt. Kephart). The fruit ripens in September when it may be harvested by red squirrels.

Baldwin [Dixon, 1961] reported a mountain winterberry from Great Smoky Mountains National Park that measured 14 inches in circumference, but in the meantime I located one at 4350 ft. near Spence Field Bald that measures 21 inches in circumference, this being the largest specimen recorded to date. This tree is approximately 25 ft. high and has a spread of about 15 ft. It grows close to the old Bote Mountain road.

This holly grows in many of the high-altitude laurel "slicks" (heath balds). It frequently sends up a dozen or more stems from the ground, these being quite uniform in size. The leaves turn yellow in color before they fall.

3. AMERICAN HOLLY—*Ilex opaca* Ait.

This evergreen tree whose thick spiny-margined leaves and red fruits are symbolic of the Christmas season is fairly common in moist situations at the low and middle altitudes. Shanks [1954a] lists it as a small tree of the cove hardwoods and a characteristic tree in hemlock forests. Whittaker [1956] who includes this holly as a member of his mesic tree class gives 4000 ft. as its uppermost limit, but in a few places such as on Spence Field Bald (4900 ft.) and Russell Field (4250 ft.) it exceeds that elevation.

Ordinarily its peak of blossoming is during the latter half of May and in early June; flowers have been noted as early as April 27 (1945) and as late as June 13 (1951). It is not unusual to find fruits of the previous year persisting on some trees at the time of flowering since the red fruits may remain until early June. Birds observed feeding upon the fruits include the pileated woodpecker, mockingbird, cedar waxwing, and various thrushes.

A number of large American hollies grow near the Meigs Creek Trail approximately 3 miles from its beginning at the Sinks on Little River. In 1958 Harold L. Hoffman's measurements of the circumference of the four largest specimens was 67, 55½, 47½, and 47 inches. Near the beginning of the trail from Cades Cove to Abrams Falls grows an American holly with circumference of 66 inches. Late in 1934 Willis King and R. J. Fleetwood measured one on the Middle Prong of Little River that was 88 inches in circumference, but this specimen disappeared during the logging operations that were still going on in that watershed.

FAMILY ✦ *CELASTRACEAE*

Euonymus L.

1. STRAWBERRY-BUSH; "HEARTS A-BUSTIN' WITH LOVE"—*Euonymus americanus* L.

This shrub is fairly common in the park where it usually grows along streams and in moist situations at low and middle altitudes. Shanks [1954a] lists it as a shrub of the cove hardwood forests, while Whittaker [1956] states that this species along with the spicebush *(Lindera benzoin)* comprises the "low-elevation mesic shrub union." It has been recorded from an altitude of 3500 ft. in the Cosby area which appears to be its upper limit in these mountains.

The usual period of flowering is during May, with blossoms noted as early as April 13 (1945) and as late as June 6 (1947). In this shrub the fruits are appreciably more showy than the flowers and account for the local name "hearts a-bustin' with love" or "hearts a-bustin.' " The capsules begin to color early in August, and by late that month they are an attractive coral-pink; in September these three- to five-lobed fruits split open exposing the bright glossy-red or orange seeds.

2. WAHOO; BURNING-BUSH—*Euonymus atropurpureus* Jacq.

On September 22, 1948, I came upon a few of these shrubs at approximately 1700 ft. in the Sugarlands where they appeared to be indigenous. The plants bore colored fruits at the time. This is the only location in the park where this rare shrub has been found. Unlike the two other species of *Euonymus,* the capsules of *E. atropurpureus* are smooth.

3. RUNNING STRAWBERRY-BUSH—*Euonymus obovatus* Nutt.

This trailing shrub is fairly common, growing mostly at middle altitudes. Shanks [1954a] listed it as a woody vine of the cove hardwoods and a ground-cover plant of the hemlock forests. It usually grows in rich rocky soils in the shade of the forest. Cain recorded it from near Indian Gap (5266 ft.)—the highest altitude at which it is known to grow. May is the usual time of flowering.

BITTERSWEET—*Celastrus* L.

1. ORIENTAL BITTERSWEET—*Celastrus orbiculatus* Thunb.

There is but one locality where this exotic vine is known to grow—along Panther Branch, near Gatlinburg, where a large number of non-native species were introduced in pre-park days.

2. CLIMBING BITTERSWEET—*Celastrus scandens* L.

Although there is no place within the park where this vine is known to occur at present, it had been reported from Cades Cove "near a limestone outcrop" by Fleetwood [1935], but quarrying operations may have served to eliminate it since subsequent search failed to locate the plant. At the north end of Fontana Dam, within a stone's throw of the park boundary, it grows profusely. Sharp recorded it from above Walland, along Little River.

The climbing bittersweet at the Fontana Dam site was in ripe fruit on October 17 (1962).

FAMILY ✦ *STAPHYLEACEAE*

BLADDERNUT—*Staphylea* L.

AMERICAN BLADDERNUT—*Staphylea trifolia* L.

This is a rare shrub in the park, its distribution here being limited to Cades Cove and vicinity. Sharp et al. [1960] report it "principally in limestone outcrops," and Jennison [1939a] restricted it to rich woods loam. Flowers have been recorded from March 31 (1935) to May 4 (1937).

FAMILY ✦ *ACERACEAE*

MAPLE—*Acer* L.

1. BOXELDER—*Acer negundo* L.

At the lower altitudes, especially in moist soils and along stream-sides, this small maple is not uncommon. Although largely confined to elevations below 2000 ft., it is known to grow at 3400 ft. along the transmountain road (Tennessee—Chimneys Parking Area) where it appears to be at its highest altitude. Its peak of bloom is during the first half of April. The fruits are a favorite item of evening grosbeaks and are eaten by various other birds.

This is the only native maple with compound leaves, the three outermost leaflets having a close resemblance to the leaves of poison ivy.

2. STRIPED MAPLE—*Acer pensylvanicum* L.

This is a fairly common tree mostly at low and middle altitudes. Its occurrence in a variety of habitats is indicated by the fact that Shanks [1954a] lists it as one of the characteristic small trees in cove hardwoods, northern hardwoods, spruce-fir, hemlock, and closed oak forests. Whittaker [1956] includes the striped maple in his submesic tree class. From the lowest altitudes in the park it ranges up to 5800 ft. (near Andrews Bald), but ordinarily it is rather scarce above 5000 ft.

The latter half of April and the early part of May is its usual time of flowering at lower altitudes, but blossoms have been recorded from March 30 (1938) to June 8 (1939)—the latter date at a high elevation.

Ordinarily this is a rather small tree, the largest specimen known to occur in the park (head of Texas Creek) measuring 39 inches in circumference.* The large three-lobed leaves, green twigs, and thin chalky-white lines in the bark help to identify this maple.

3. NORWAY MAPLE—*Acer platanoides* L.

Shanks [1961] reported one of these European trees growing near an old hotel site in Greenbrier Cove. This appears to be the only locality in the park where this exotic maple was introduced.

4. RED MAPLE†—*Acer rubrum* L.

This is a very common to abundant tree throughout lower and middle altitudes and fairly common up to 6000 ft. Shanks [1954a] lists it as a major species in all but the spruce-fir forest type where it is regarded as an occasional tree. Whittaker [1956] lists it as one of the species comprising his submesic tree class. I noted a specimen of this maple at 6150 ft. under Myrtle Point of Mt. Le Conte on September 22 (1943) when its leaves were in full autumn color; this appears to be the highest altitude at which it grows in the park.

Flowers of red maple usually appear in February, although there are records of January blossoms in 1937, 1949, and 1950—the earliest being January 14 (1950). At high altitudes this species may still have flowers at the beginning of May (1947, 1958).

The largest specimen in the park, discovered near an old man-way along Lowes Creek (Greenbrier area), measures 17 ft. 3 inches in circumference (July, 1959).‡

5. SILVER MAPLE—*Acer saccharinum* L.

This tree has been introduced in a few localities near the park boundary. Along Route 73, about 2 miles east of Gatlinburg, a number of these fast-growing maples are to be found at a house site within a stone's throw of the park line. Along this same road near Baxter's Orchard (vicinity of

* According to American Forestry Association records [Dixon, 1961], the largest striped maple, 46 inches in circumference, grows near Princeton, Mass.

† The name *Acer rubrum* var. *trilobum* K. Koch has been applied to the red maples with smaller more compact leaves that grow in parts of Cades Cove (gum swamp) and near the Oconaluftee Visitor Center. This variety is not accepted by Little [1953].

‡ The largest recorded specimen, 20 ft. 8½ inches in curcumference, grows at Annandale, N. Y. [Dixon, 1961].

Cosby) other specimens grow near the park. The silver maple has also been planted in Gatlinburg.

6. SUGAR MAPLE—*Acer saccharum* Marsh.

This is a common forest tree in rich soils at low and middle altitudes. According to Shanks [1954a], it is a dominant species in cove hardwoods, northern hardwoods, and in hemlock forests. Whittaker [1956] places it in his mesic tree class. Ayres and Ashe [1905] estimated that 10 percent of the forest cover in the Little River watershed consisted of sugar maples in pre-park days.

This maple becomes quite scarce above 5000 ft. Whittaker [1956] indicates it may extend to 5700 ft. in these mountains. Flowers have been noted as early as March 16 (1953, near Park Headquarters) and as late as May 16 (1940, near Spence Field Bald). The fruits, unlike those of the red maple, mature in the autumn.

The largest sugar maple recorded within the park grows at an altitude of 3400 ft. along the trail from Cherokee Orchard to Trillium Gap; it measures 15 ft. 4 inches in circumference.* There are a number of fine specimens whose circumference exceeds 13 ft.

The Sugarlands, a valley approximately 4 miles long extending along the West Prong of the Little Pigeon River from the Sugarlands Visitor Center to Chimneys Campground, gets its name from the large sugar maples that grew there in pre-park days. These trees were usually tapped for their sweet sap in late winter.

7. MOUNTAIN MAPLE—*Acer spicatum* Lam.

In rich rocky soils at the higher altitudes this small tree is quite common, usually growing in the shade of other trees. Shanks [1954a] lists it as a small tree of cove hardwoods, northern hardwoods, and spruce-fir forests. Whittaker [1956] groups it with yellow birch, Allegheny serviceberry, and alternate-leaf dogwood in an "ecotonal-mesic union, centered in mesic sites at elevations around 4500 ft." From near the summits of the highest mountains it may extend down to approximately 3000 ft., such as along the

* According to the American Forestry Association, the largest specimen of sugar maple, 19 ft. 9 inches in circumference, grows in Garrett County, Md. [Dixon, 1961].

Big Locust Nature Trail (Chimneys Campground area) and on Bradley Fork. The lowest elevation at which it occurs is 2250 ft. at the old Russell Bohanan place along Wooly Tops Branch (Greenbrier area).

Early June is the normal time of flowering of the mountain maple. Earliest blossoms were noted May 8 (1948) at 3700 ft., while on July 11 (1940) the tree was going out of flower on Mt. Kephart.

This small tree often assumes a shrubby type of growth. A specimen discovered by S. G. Baldwin along the Boulevard Trail (Mt. Le Conte) has a circumference of 3 ft., an estimated height of 25 ft., and a spread of 10 ft.; this appears to be the largest specimen of mountain maple on record [Dixon, 1961].

FAMILY + *HIPPOCASTANACEAE*

BUCKEYE—*Aesculus* L.

YELLOW BUCKEYE—*Aesculus octandra* Marsh.

In coves where soils are deep and rich this is a common tree. Shanks [1954a] listed it as dominant in cove hardwoods and northern hardwoods and an occasional tree in spruce-fir forests. Ayres and Ashe [1905] estimated that in pre-park days this species comprised 15 percent of the forest cover in the bottoms of the Cataloochee district. Whittaker [1956] places the yellow buckeye in his mesic tree class. From near the lowest altitudes in the park (900 ft. along lower Abrams Creek) the range of this tree extends upward to 6300 ft. (Clingmans Dome Parking Area).

The latter half of April and all of May represent the usual time of flowering, depending much on altitude. Blossoms have been noted as early as March 30 (1938) and as late as June 13 (1947, 1948). Since cattle and horses are poisoned by eating the glossy brown fruits, in pre-park days these animals were driven from those grass balds where buckeyes grew (e.g., Spence Field) prior to the ripening and dropping of the fruits. It is not known whether the red and the gray squirrels, both of which eat large quantities of buckeyes, are affected in any way.

Within a few years after the American chestnut succumbs to the relentless fungus disease, the bark sloughs off. Some of these dead forest giants may remain standing for over a quarter of a century.

Early summer would mark the peak of bloom of the ill-fated American chestnut *(Castanea dentata)* at middle and high altitudes. All of the mature trees have yielded to the ravages of the oriental blight.

Trail through cove hardwood forest near Ramsey Prong. Trees are yellow birch *(Betula alleghaniensis)* on left, eastern hemlocks *(Tsuga canadensis)* center, and yellow-poplar *(Liriodendron tulipifera)* on right. The understory is comprised largely of rosebay rhododendron *(Rhododendron maximum).*

This, the largest eastern hemlock *(Tsuga canadensis)* on record—19 ft. 10 inches in circumference, grows on the north slope of Mt. Le Conte.

Yellow-poplars *(Liriodendron tulipifera)* measuring over 5 ft. in diameter are more prevalent in the Great Smoky Mountains National Park than in any other area.

The hazel alder *(Alnus serrulata)* is ordinarily the first woody plant to come into flower. In mild winters this shrub will display its straw-colored tassels in January.

Witch-hazel *(Hamamelis virginiana),* the last of the woody plants to come into flower, displays its thin yellow-colored petals throughout October and November.

Baldwin [1948b] reports the finding of a yellow buckeye 15 ft. 11 inches in circumference near the trail from Cherokee Orchard to Trillium Gap—the largest of its kind ever recorded. (A photograph of this tree is one of the illustrations in his article.) I measured a yellow buckeye of 14 ft. circumference approximately ¼ mile below Mill Creek Falls in the Cades Cove area. Another, just south of the Chimney Tops, is 15 ft. 9 inches in circumference and approximately 100 ft. high. Many other big buckeyes grow in the park.

This is the "only tree in the park with palmately-compound five-foliate leaves" [Sharp, 1942a]. Jennison [1939a] called attention to the fact that the buckeye one finds in exposed sites at high altitudes (e.g., Spence Field Bald) is quite dwarfed in appearance; there "it propagates and increases its range by means of root sprouts."

Excellent specimens of this large forest tree grow along the trail to the Chimney Tops, the Buckeye Nature Trail, and elsewhere. There are no other species of buckeye in the park, although both the Ohio buckeye *(A. glabra)* and the red buckeye *(A. pavia)* occur in the vicinity.

FAMILY ✦ *RHAMNACEAE*

BUCKTHORN—*Rhamnus* L.

CAROLINA BUCKTHORN—*Rhamnus caroliniana* Walt.

Although there are a few localities where it grows quite abundantly, this shrub is rare in the park. Its best growth is in the Hearon field along the old Cooper road, approximately 1 mile from the Abrams Creek Campground. It also grows in Cades Cove and along Laurel Creek within the park and in Dry Valley, Wears Cove, Gatlinburg, and near Laurel Lake close to the boundary. All these localities are at low altitudes (below 2000 ft.). It appears to grow best on limestone soils.

Flowers have been noted during June. Sharp [1942a] writes that the twigs "are yellowish and ill-smelling when crushed" and that this species is "closely related to the tree from which is derived the drug, cascara."

Ceanothus L.

NEW JERSEY TEA—*Ceanothus americanus* L.

This is a common shrub at low and middle altitudes "in poor soils on banks and open wooded slopes" [Jennison, 1939a]. Shanks [1954a] lists it as one of the lower shrubs in open oak and pine stands. Whittaker [1956] includes it as a species in his sub-mesic shrub union. From the lowest altitudes it extends upward to 5462 ft. (Soco Bald) on the southeastern boundary.

June is the usual time of flowering. The earliest blossoms were noted along lower Abrams Creek on May 18 (1948) and the latest on July 21 (1939).

FAMILY ✦ *VITACEAE*

Ampelopsis Michx.

HEARTLEAF AMPELOPSIS—*Ampelopsis cordata* Michx.

There is but one record for this rare woody vine in the park. Sharp discovered a specimen near the mouth of Abrams Creek (elevation approximately 900 ft.) in October 1956.

Parthenocissus Planch.

VIRGINIA CREEPER*—*Parthenocissus quinquefolia* (L.) Planch.

This high-climbing woody vine is common at low and middle altitudes. According to Shanks [1954a], it is a plant of cove hardwoods and closed oak forests. Flowers have been noted from middle June to middle July. The fruits resemble a cluster of small grapes. In the autumn the leaves of the Virginia creeper color a brilliant red.

* The variety *hirsuta* (Donn) Fern. was reported by Sharp [1942b] as "rare, similar to the species except more hairy." Jennison [1939a] also listed this variety as a park entity.

GRAPE—*Vitis* L.

1. SUMMER GRAPE*—*Vitis aestivalis* Michx.

This grapevine is "frequent in sandy soils, thickets, and rocky woodlands" [Jennison, 1939a]. Dr. Magoon reported it to me from along Roaring Fork, and Shanks [1954a] calls it a plant of the cove hardwood forest. This and the variety *argentifolia* are quite similar in appearance, a matter that is discussed at some length by Bailey [1934a].

2. POSSUM GRAPE†—*Vitis baileyana* Munson

This woody vine is a Southern Appalachian species occurring on "moist sandy loams on wooded slopes and along streams at lower elevations" [Jennison, 1939a]. It appears to be uncommon, growing mostly in the western end of the park (Abrams Creek, Tab Cat Creek).

3. PIGEON GRAPE—*Vitis cinerea* Engelm.

Not much is known about the status of this woody vine which, according to Jennison [1939a], grows "in woods loam and sandy alluvia at lower elevations." It is recorded from the park by Sharp et al. [1960] and by Shanks [1961].

4. FOX GRAPE—*Vitis labrusca* L.

This woody vine, characterized by dense rusty-colored hairs covering the undersurfaces of the leaves and by its large (½ inch diameter) fruits, is rare in the park. Sharp [1942b] reported it from

* The variety *argentifolia* (Munson) Fern., formerly known as *V. argentifolia* Munson, is called "frequent and locally abundant" by Jennison [1939a] who writes that it occurs "in well drained sandy loams on rocky woodland slopes, mostly at lower and middle elevations." Following a visit to the park in 1941, Dr. C. A. Magoon informed me (letter) that this is the vine "showing such abundant and rampant growth—between your headquarters and the 'Loop,' and also along the upper Fighting Creek portion of Highway 73." Gilbert [1954] reported it from 4700 ft. on Parson Bald which may be near its highest altitude in the park. Flowers have been noted in late May.

† Gleason [1952] writes that it is a "problematical species combining the general characters of *V. cordifolia* [*V. vulpina*] with the pubescens of *V. aestivalis*."

Indian Creek (tributary of Deep Creek) and from near Nellie (Cataloochee area), N. C.

5. MUSCADINE—*Vitis rotundifolia* Michx.

Below 2500 ft. where it normally grows along streamcourses and in moist soils this woody vine is fairly common. June is the usual time of flowering and the fruits ripen in September and October. This grape is readily distinguished from other native grapes by its large tough-skinned fruits of which there are but few in a cluster. The fruits are not persistent on the vines like in many other species. The leaves are much smaller than other grape leaves, very glossy above and below, and margined by large teeth.

6. FROST GRAPE—*Vitis vulpina* L.

This high-climbing woody vine has been noted in a number of localities—all of which are at low elevations. An unusually large specimen of what appears to be this species was reported by H. S. Pepoon (Chicago) who called it the "chicken grapevine" *(Vitis cordifolia).** This specimen, growing near Dunns Creek (Cosby area), measured 60½ inches in circumference.

FAMILY ✦ *TILIACEAE*

BASSWOOD—*Tilia* L.

In the various manuals there appears to be little or no agreement as to the taxonomy of the genus *Tilia*. In their *Trees of the Southeastern States,* Coker and Totten [1934] include 13 native species of *Tilia* in addition to some ornamentals. The present tendency is to regard most of the so-called species as variants. In "Louisiana Trees and Shrubs" [Brown, 1945] the author writes: "The reduction of 7 species and varieties reported for Louisiana to one is not due to ultra-conservatism, but is the result of the study of a series of specimens. . . . This study indicates that degree of hairiness has little or no value."

* Letter dated March 16, 1935, with which was enclosed a drawing of the vine and a detail of the leaf.

The University of Tennessee Herbarium has 27 specimens (sheets) of the genus *Tilia* from Great Smoky Mountains National Park. These had been identified as comprising seven species until 1958 when an examination by G. N. Jones (Univ. Illinois) reduced them to one: *T. heterophylla* Vent.

WHITE BASSWOOD—*Tilia heterophylla* Vent.

This is one of the dominant trees of the cove hardwood forests [Shanks, 1954a]; elsewhere it grows along stream banks and in moist soils to an elevation of approximately 5000 ft. In pre-park days, according to Ayres and Ashe [1905], the "linn" (contraction of "linden") comprised 12 percent of the forest in the Little River watershed below the Elkmont turnoff. It is one of the species in Whittaker's mesic tree class.

Flowering usually occurs from middle June to middle July, at which time these trees become alive with bees. The ground beneath the basswoods becomes littered with the pea-like fruits in late July and in August.

Along the Big Locust Nature Trail (Chimneys Campground area) is the stump of a basswood that measures 12 ft. in circumference.* A basswood growing along the Porters Creek Trail (Greenbrier area) measures 11 ft. 7 inches in circumference.

When the breezes are strong enough to stir its leaves, a basswood may be identified at a considerable distance by the gray or silvery undersurface of its leaves. Ordinarily this tree sends up a number of sprouts from the base; occasionally it becomes ringed with basal shoots.

FAMILY + *MALVACEAE*

Hibiscus L.

1. SWAMP ROSE-MALLOW†—*Hibiscus palustris* L.

This shrub is rare in the park. Sharp is of the opinion that the

---

* According to the American Forestry Association [Dixon, 1961], the largest recorded specimen measures 13 ft. 9 inches in circumference; it is located near Little Switzerland, N. C.

† The variety *oculiroseus* is indicated on specimens in the University of Tennessee Herbarium.

specimens in the wet Oliver meadow in Cades Cove may be regarded as indigenous whereas the plant found in the Fighting Creek valley was probably introduced. The plants in Cades Cove were in flower on August 17 (1946).

2. ROSE-OF-SHARON—*Hibiscus syriacus* L.

There appears to be but one record of this exotic shrub—a specimen along the Pink Root Trail near Laurel Creek. It was in flower on July 16 (1937). The rose-of-Sharon was probably introduced around homesites in other parts of the park.

FAMILY + *THEACEAE*

Stewartia L.

MOUNTAIN STEWARTIA—*Stewartia ovata* (Cav.) Weatherby

This plant, ordinarily a shrub, is rather uncommon. It grows at altitudes between 1000 and 2500 ft. "in moist sandy loam on wooded banks and along streams" [Jennison, 1939a]. On the North Carolina side of the park it has been recorded only from Twentymile Creek, and it grows nearby in Fontana Village where it is indigenous. On the Tennessee side it occurs in a number of localities: Abrams Creek, Abrams Falls, Laurel Creek, Parson Branch, Middle Prong of Little River, Sinks on Little River, Sugarlands, Gatlinburg (Cherokee Orchard Road, Baskins Creek, and Roaring Fork), and near Greenbrier.

The large white or cream-colored flowers have numerous yellow or purple stamens. They are usually at their height of blossoming during the last half of June, the extreme dates being June 4 (1945) and July 16 (1951). The mountain stewartia is one of the endemic plants of the Southern Appalachians and the only representative of the tea family in the park.

FAMILY ✦ *GUTTIFERAE*

Ascyrum L.

ST. ANDREW'S CROSS*—*Ascyrum hypericoides* L.

This dwarf shrub is not uncommon in the park, where it grows in sandy or dry situations mostly below 3000 ft. There is one record of its occurrence at 4250 ft. (Russell Field). Its period of flowering is quite extensive, covering a span of three months beginning in early July. The flowers have four yellow petals arranged in the form of an "X."

FAMILY ✦ *NYSSACEAE*

TUPELO—*Nyssa* L.

BLACKGUM—*Nyssa sylvatica* Marsh.

This is a common tree of low and middle altitudes where it grows in fairly dry exposed places. Shanks [1954a] lists it as a non-dominant species in cove hardwoods and a dominant tree in closed oak forests and in open oak and pine stands. It is in the subxeric tree class of Whittaker [1956] who gives its upper limit as 5300 ft.

Ordinarily its peak of flowering is early May, with April 11 (1945) and May 29 (1937) as the extreme dates. I have measured specimens up to 10 ft. 7 inches in circumference (Baskins Creek).†

The glossy blackgum foliage takes on a blood-red hue in the autumn, adding materially to the attractiveness of the mountainsides wherever it grows. The small dark fruits are eaten by bears, squirrels, and various birds.

---

* Among the specimens of *Ascyrum hypericoides* in the herbarium of the University of Tennessee are 12 that were collected in Great Smoky Mountains National Park. William P. Adams, who examined these in 1956, indicated that four should be called *A.h. hypericoides* and eight should be called *A.h. multicaule*.

† According to the American Forestry Association [Dixon, 1961], the largest recorded specimen measures 13 ft. 3 inches in circumference; it is located at Sandy Spring, Md.

FAMILY **+ *ARALIACEAE***

Aralia L.

DEVILS-WALKINGSTICK; HERCULES-CLUB—
Aralia spinosa L.

This stout-stemmed plant, usually armed with a formidable array of sharp prickles, may be regarded as either a shrub or a small tree. It grows in a variety of situations, mostly at low and middle altitudes, where its tendency to sucker often results in a small dense grove of these plants. Ordinarily its range is under 3600 ft. altitude, but it grows at 4200 ft. near the old Bote Mountain road, and I counted a grove of 17 specimens along the Appalachian Trail on Spence Field Bald at 5000 ft. Shanks [1954a] lists it as a small tree of the cove hardwood forest, and Whittaker [1956] includes it as a species in his submesic shrub union.

The latter half of July usually marks the peak time of flowering. It may come into blossom as early as July 3 (1945), and there is a small grove of these trees at 3600 ft. beside the trans-mountain road (Tennessee side) that flower throughout September —at a time when this species bears ripe fruit at lower elevations. The fruits are eaten by many kinds of birds.

A number of years ago Baldwin reported a specimen of devils-walkingstick from 3100 ft. along the Ramsey Cascades Trail that was 25 inches in circumference [Dixon, 1961]. In January 1961 I measured one that was 27 inches in circumference—at Wild Cherry Hollow (3500 ft.) near the main transmountain road (North Carolina side). This latter record specimen has an estimated height and spread of 26 and 20 ft., respectively.

FAMILY **+ *CORNACEAE***

DOGWOOD—*Cornus* L.

1. ALTERNATE-LEAF DOGWOOD—*Cornus alternifolia* L. f.

This is a rather uncommon shrub (or small tree) that grows in fairly rich soils from 1100 ft. upward to near the summits of the highest mountains (6500 ft. on Clingmans Dome). There are but few woody plants in the park with so extensive a range in altitude. Shanks [1954a] lists it as a small tree of the cove hardwoods, northern hardwoods, hemlock forests, and spruce-fir forests. It is one of the species (with yellow birch, mountain maple, and Allegheny serviceberry) that Whittaker [1956] puts in his "ecotonal-mesic union, centered in mesic sites at elevations around 4500 ft."

Flowering takes place mostly in May and June, the extreme dates being April 16 (1945, at 1200 ft.) and July 11 (1940, about 6000 ft.).

A specimen with circumference of 12 inches and height estimated as 30 ft. grows at 4250 ft. near the Gregory Ridge Trail. A shorter specimen with circumference of 14½ inches grows near the main transmountain road (North Carolina side) at 4200 ft.*

2. SILKY DOGWOOD—*Cornus amomum* Mill.

Although it is quite common in a number of low-altitude localities, this shrub is scarce or absent over most of the park. Except for its occurrence in the Cades Cove and Park Headquarters areas it grows largely on or near the boundary, especially at the western (Abrams Creek, Happy Valley) and southern (Fontana Reservoir, Deep Creek, Noland Creek) limits. The silky dogwood grows in or near water or in very wet soils from the lowest altitudes to 2000-2500 ft.

Flowers have been recorded from late May to late June. The mature fruits are pale blue in color in late August and in September. The presence of rusty-brown hairs on the undersurface of the leaves will serve to separate this from the other two dogwoods in the area.

3. FLOWERING DOGWOOD—*Cornus florida* L.

Below 3000 ft. this is one of the most abundant small trees in the park. Shanks [1954a] listed it as a small tree of the cove hardwood and closed oak forests. It is one of the species in Whittaker's [1956] submesic tree class. Although it usually will drop out rather rapidly at altitudes above 3000 ft., specimens may occur up to near 5000 ft. (Spence Field Bald).

* The largest specimen recorded by the American Forestry Association [Dixon, 1961] measures 21 inches in circumference; it grows at Rupert, Vt.

Normally the peak of flowering takes place during the second, third, or fourth week of April, depending on the climatic conditions that prevailed prior to those times. Some blossoms have been recorded as early as March 17 (1945) at a low altitude and as late as June 8 (1960) at 4600 ft. In most years the peak of flowering has passed by the end of April. A few flowers are often reported in autumn, but these are mostly stunted or otherwise abnormal. The bright glossy red fruits are a source of food for many kinds of birds and for squirrels and other wildlife.

What appears to be the largest flowering dogwood in the park, measuring 4 ft. 5½ inches in circumference,* was reported growing along Bradley Fork, approximately 1¾ miles above the Smokemont Campground.

FAMILY ✦ *CLETHRACEAE*

Clethra L.

CINNAMON CLETHRA—*Clethra acuminata* Michx.

This plant, ordinarily a tall shrub, is of fairly common occurrence "in well drained soil in woods" [Jennison, 1939a]. Shanks [1954a] lists it as a shrub of the cove hardwoods, closed oak forests, and open oak and pine stands. It represents one of the species in Whittaker's [1956] submesic tree class. In elevation it ranges from near 1500 ft. (the Sinks on Little River) to 5500 ft. (Rocky Spur on Mt. Le Conte).

Ordinarily the cinnamon clethra is at its peak of flowering during the latter half of July, the extreme dates being June 23 (1945, at 2600 ft.) and August 24 (1939, near 5500 ft.).

A specimen 15 ft. high with a trunk diameter of 3 inches was measured at 4300 ft. between Cosby Knob and Mt. Cammerer.

* The largest specimen reported to the American Forestry Association, measuring 5 ft. 4 inches in circumference, grows near Oriole, Md. [Dixon, 1961]. Coker and Totten [1934] mention two from near the northeastern corner of North Carolina: one measures 2 ft. 2 1/5 inches in diameter at 2 feet from the ground, and the other measures 1 ft. 10 1/5 inches in diameter at breast height.

The thin reddish-brown bark that is constantly exfoliating is characteristic of the more mature plants.

The cinnamon clethra is one of the endemic plants of the Southern Appalachian Mountains.

FAMILY ✦ *PYROLACEAE*

Chimaphila Pursh

SPOTTED WINTERGREEN—*Chimaphila maculata* (L.) Pursh

This low-growing semi-herbaceous plant is common in dry woodlands, especially in acid soil under pines. It grows from the lowest altitudes to 4450 ft. (Noland Divide Trail near Coburn Knob).*
Shanks [1954a] includes it in a list of herbaceous ground-cover plants of closed oak forests.

The peak of flowering is usually during the latter half of June, with extreme dates of June 7 (1959) and July 14 (1937).

FAMILY ✦ *ERICACEAE*

Rhododendron L.

1. SMOOTH AZALEA†—*Rhododendron arborescens*
   (Pursh) Torr.

This Southern Appalachian endemic is uncommon and localized. Except for its occurrence on Gregory Bald at 4900 ft. where it is one of the species involved in the fantastic variety of natural hybrid azaleas (see flame azalea, *R. calendulaceum),* it grows mostly at low altitudes along streamcourses. It occurs along Little River between Millsap Branch and the Sinks, along lower Abrams Creek, near Abrams Falls, and along the West Fork of the Little

* On June 29, 1935, Jennison and L. Barksdale collected a specimen "near Andrews Bald," but gave no altitude.

† According to Sharp et al. [1960] and Shanks [1961], the variety *richardsonii* Rehd. is also represented.

Pigeon River about 2 miles below Gatlinburg. A white-flowered azalea reported growing in the Cataloochee area along Mill Creek has not been examined.

Middle June is the usual time of flowering with extreme dates of May 23 (1943) and July 3 (1935). The flowers are exceedingly fragrant and the corolla tubes very sticky.

2. FLAME AZALEA—*Rhododendron calendulaceum* (Michx.) Torr.

This fine shrub is a common plant of the closed oak forests and of the open oak and pine stands [Shanks, 1954a]. It grows from the lowest altitudes to 5800 ft. (Andrews Bald). The flame azalea is one of the species making up the submesic shrub union of Whittaker [1956]. Its normal time of flowering is from late April at the lower altitudes to early July near the upper limits of its range. I observed blossoms on flame azaleas at 2000 ft. near Fighting Creek Gap on March 30, 1945, near the close of the warmest March on record. At the higher altitudes a few flowers have been noted through the summer months and on as late as September 25 (1938).

William Bartram [1791], pioneer plant explorer, described "this most celebrated species of Azalea" as being

> in general of the colour of the finest red lead, orange and bright gold, as well as yellow and cream colour; these various splendid colours are not only in separate plants, but frequently all the varieties and shades are seen in separate branches on the same plant; and the clusters of the blossoms cover the shrubs in such incredible profusion of the hillsides that suddenly opening to view from dark shades, we are alarmed with apprehension of the hill being set on fire. This is certainly the most gay and brilliant flowering shrub yet known.

Bartram's remarks would appear to be particularly appropriate to the azalea situation prevailing on Gregory Bald (4948 ft.) where there are hundreds of native azaleas whose colors range "in every imaginable hue from pure white to pale yellow, salmon yellow, clear pink and orange-red to red" [Skinner, 1955], except for the fact that botanists [Camp, 1936; Skinner, 1955; Galle, 1963] are inclined to the belief that of the three species involved in the hybridization process at that place, the flame azalea is represented only as "a late blooming variety," "a hybrid," "an ancestral type," "a relative," "a subspecies," etc. They are in agreement, however, with the occurrence of the true flame azalea along the trails leading up to Gregory Bald. Some of these plants grow to a height of 18 ft. (Hannah Mountain) in the forest.

Flame azaleas grow in considerable numbers along the Heintooga Overlook spur from the Blue Ridge Parkway where the usual peak of blossoming takes place during the first half of June. The Gregory Bald display, where Camp [1936] found "the most bizarre and taxonomically most vexing natural collection of azaleas," will reach its height of flowering between the middle and the end of June, depending on the progress of the season. On Andrews Bald, where there are many fine flame azaleas, the flowering peak is ordinarily in middle June. Elsewhere, for the most part, the flame azaleas are quite scattered, and over large areas of the park this shrub is rather scarce or absent.

3. PIEDMONT AZALEA—*Rhododendron canescens* (Michx.) Sweet

This pink-flowered shrub appears to be restricted to a few low-altitude localities in the extreme western end of the park and vicinity. It has been reported on Revenue Hill, near Deals Gap, and on Chilhowee Mountain and in the Calderwood area. The attractive flowers appear before the leaves, usually between late April and middle May.

In July, 1951, when Dr. Henry T. Skinner called at my office and examined specimens of azaleas in the park herbarium, it was his opinion that the pink-flowered plants from the western boundary and vicinity should be called *R. canescens*. According to Dr. A. J. Sharp (1963), *R. nudiflorum* also grows in that area.

4. PINXTER-FLOWER—*Rhododendron nudiflorum* (L.) Torr.

In view of the statements made pertaining to the similar-appearing piedmont azalea *(R. canescens),* the pinxter-flower is mentioned here in view of Dr. Sharp's remark that it occurs in the vicinity of the park's western boundary. Further study of these pink-flowered azaleas is necessary.

5. SWAMP AZALEA*—*Rhododendron viscosum* (L.) Torr.

This fragrant white- or pink-flowered shrub appears to be restricted to Parson and Gregory Balds on the state line in the southwestern part of the park. Skinner [1955] found it on Parson Bald in 1951 in his quest for species of azaleas that were involved in the hybridization that resulted in the bewildering variety of flower-colors on nearby Gregory Bald. He describes it as small with sticky flowers and a suckering root system. "Certainly it, with the other two species mentioned [*R. calendulaceum* in modified form and *R. arborescens*], was involved as an original parent of these plants" [ibid.].

Flowers have been noted in late June and early July.

6. CATAWBA RHODODENDRON—
Rhododendron catawbiense Michx.

Whereas the preceding species of the genus *Rhododendron* are deciduous and go by the common name of "azaleas," the present species along with those that follow it are evergreen and may be regarded as true "rhododendrons." Of the latter group the Catawba, with its profusion of purplish-pink blossoms, is, when in flower, the most spectacular and the greatest favorite with the visiting public. As far as questions pertaining to plants are concerned, certainly one of the most common is "When do the rhododendrons bloom?"—the usual reference being to *R. catawbiense*.

Shanks [1954a] regarded this Southern Appalachian endemic as a shrub of the hemlock, northern hardwood, and spruce-fir forests. It is abundant on many high-altitude (above 4000 ft.) ridges where it is one of the major components of the heath balds ("laurel slicks"). Cain [1930b] concluded that of the three catastrophic factors influencing the formation and maintenance of heath balds in these mountains—windfall, landslide, and fire— "fire is by far the most important."

Along the main transmountain road (Tennessee side) the lowest occurrence of the Catawba rhododendron is at 3500 ft. (between Chimneys Parking Area and the tunnel overpass). Along the Bull Head Trail to Mt. Le Conte this shrub drops down to 2700 ft. Occasionally one finds one or a few specimens as low as

---

* The variety *montanum* Rehd. is represented according to determinations by W. H. Camp and the writing of Skinner [1955].

1600 ft. as in the Buckhorn area near Emerts Cove. This shrub has its uppermost range at about 6550 ft. on Mt. Le Conte.

Ordinarily the Catawba rhododendrons in the vicinity of Alum Cave Bluffs are at their peak of flowering June 5-15; this is one of the best locations for viewing these plants since the trail not only penetrates a heath bald where these shrubs are dominant, but at this bald (4600-4700 ft.) the views of the nearby rugged mountains (Huggins Hell and Anakeesta, in particular) are spectacular. At higher altitudes on all park trails, especially at 5500-6000 ft., the blossoming peak is likely to be June 20-25. The earliest flowers were recorded on May 8 (1945) near the tunnel overpass and on May 8 (1948) along the Bull Head Trail (2700-4200 ft.). A few flowers were noted as late as July 17 (1940) near the summit of Mt. Le Conte, and occasionally a few appear in August, September, and October (October 16, 1941—vicinity of Alum Cave Bluffs).

This shrub does not flower heavily every year; neither does this occur on alternate years, as stated by Jennison [1939a]. One might say that during most years the amount of blossoming is quite rewarding, but there are years when an exceptionally heavy flowering takes place and other years when it is quite light. Such occurrences appear to have no regularity and are therefore unpredictable.

What appears to be a white-flowered hybrid of this and the rosebay rhododendron *(R. maximum)* grows on Andrews Bald in at least two localities where I have noted its flowers between June 14 and July 22. Ramseur [1959] mentions such a hybrid in the Black Mountains of North Carolina, and Whittaker writes of having "what seems convincing evidence of introgressive hybridization between *R. maximum* and *R. catawbiense"* (letter, 1960).

7. ROSEBAY RHODODENDRON—*Rhododendron maximum* L.

This large evergreen shrub is common throughout most of the park except at the high altitudes. Shanks [1954a] listed it as a characteristic shrub of five of the six major forest types—all except spruce-fir. It is the most abundant of all our rhododendrons. From the lowest altitudes it extends upward to approximately 5000 ft. and occasionally to 6000 ft., and there is evidence of hybridization with the Catawba rhododendron at these higher elevations (see *R. catawbiense*). The rosebay rhododendron makes its best growth along streams in the shade of the forest, in ravines and hollows, and on moist slopes. The "mesic heath union" of Whittaker [1956] consists of this shrub along with the drooping leucothoë *(Leucothoë fontanesiana).*

This rhododendron blooms mostly in June throughout the lower altitudes and in July at the middle elevations. Flowers have been noted as early as May 15 (1955); some late flowers may be found throughout August, with stray blossoms in September and early October (October 10, 1937). The flowers are ordinarily a waxy white, with greenish-yellow spots on the upper lobe; occasionally the petals are a pale rose-pink. The flower buds are rose-colored before they open.

The thick, dark, evergreen leaves react to subfreezing temperatures by rolling into a tight coil and bending downward; such a position is also assumed as a result of severe drouth. Specimens large enough to be called trees are common. This rhododendron is known to attain a trunk diameter of 9 inches and to reach a height of over 20 ft.

8. PIEDMONT RHODODENDRON—*Rhododendron minus* Michx.

On exposed ridges high in the mountains, ordinarily above 4500 ft., one may find a low-growing evergreen rhododendron whose leaves are approximately the size and shape of the leaves of mountain-laurel *(Kalmia latifolia).* The undersurfaces of the leaves of this dwarf rhododendron, however, are quite rusty in color. This shrub, which for many years was called *R. carolinianum, R. minus, R. punctatum,* and possibly by other names, grows abundantly on Cliff Top of Mt. Le Conte and in a number of other localities (Alum Cave Bluffs, Rocky Spur, Chimney Tops, Charlies Bunion, Mt. Sequoyah, and elsewhere). Its rose-colored flowers are normally on display during late June and early July, the extreme dates being June 4 (1941) and August 12 (1936).

There is quite an altitudinal gap between the occurrence of the aforementioned rhododendron and a taller-growing but otherwise similar-appearing shrub that grows along lower Noland Creek and near the mouth of Twentymile Creek on the North Carolina side of the park, and along lower Abrams Creek and Parson

Branch on the Tennessee side. The latter rhododendron grows in the forest, usually near streams, and may reach a height of 10 ft. The leaves and flowers of these two small-leaved rhododendrons are quite similar, and some botanists such as Jennison [1939a] believed that only one species was involved, the minor differences being a reflection of the very dissimilar environment. The recent study by Duncan and Pullen [1962] has confirmed Jennison's opinion, all the evergreen small-leaved rhododendrons in the park now being regarded as *R. minus*.

The Piedmont rhododendron of the lower altitudes is usually in flower from middle May to middle June.

Menziesia Sm.

MINNIE-BUSH—*Menziesia pilosa* (Michx.) Juss.

Being a shrub of the spruce-fir forests [Shanks 1954a] this plant is largely restricted to the eastern half of the park (east of Clingmans Dome) where it occurs from 5300 to 6600 ft. Whittaker [1956] lists it as a member of the "low-shrub union of subalpine forests." It grows quite commonly high up on Mt. Le Conte, and it probably occurs on all the mountains where spruce-fir prevails. The minnie-bush is endemic in the Southern Appalachian Mountains.

The cream-colored flowers resemble those of the blueberry. Usually the peak of flowering is in late June, the extreme dates being June 5 (1941) and July 23 (1942).

SAND-MYRTLE—*Leiophyllum* (Pers.) Hedw. fil.

HUGER'S SAND-MYRTLE*—*Leiophyllum buxifolium* var. *hugeri* (Small) Schneid.

This dense, low-growing, small-leaved, evergreen shrub occurs rather frequently and, occasionally, abundantly on exposed ridges and summits mostly above 4500 ft. From approximately 3700 ft.

* As a result of his study of *Leiophyllum*, Camp [1938] referred to the Great Smoky Mountains plant as *L. lyoni* Sweet. Copeland [1943] calls it *Dendrium lyoni* Sweet.

(Bear Pen Hollow) it extends upward to 6600 ft., and although it is found in spruce-fir forests for the most part, there are a few small isolated colonies well beyond the extent of spruce-fir, such as on Gregory Bald. According to Whittaker [1956], this plant is a member of the high-elevation heath bald shrub union. In addition to its occurrence on Mt. Le Conte and its various spurs (Alum Cave Bluffs, Brushy Mountain, Rocky Spur) it grows near the Jumpoff, Charlies Bunion, Chimney Tops, Maddron Bald, and elsewhere.

Flowers have been noted on Brushy Mountain and near Alum Cave Bluffs as early as April 15 (1946), but the peak of blossoming at those places is usually in late May and early June. High up on Mt. Le Conte the height of flowering is in June, with a few blossoms in evidence until middle July. A relatively light second flowering can be expected in the autumn, especially in September and early October. The latest flowers noted near the summit of Mt. Le Conte were on October 26 (1939).

Like several other ericaceous shrubs, Huger's sand-myrtle is endemic in the Southern Appalachian Mountains.

LAUREL—*Kalmia* L.

MOUNTAIN-LAUREL—*Kalmia latifolia* L.

At low and middle altitudes this is a common shrub in many areas of the park. Shanks [1954a] lists it as a shrub of cove hardwood, hemlock, and closed oak forests, and of open oak and pine stands. According to Whittaker [1956], the mountain-laurel is a member of both the subxeric heath union and of the lower-elevation heath bald shrub union. It is one of the most abundant plants of the heath balds (see *Rhododendron catawbiense)* and thrives under a wide range of conditions.

Mountain-laurel is in flower throughout May and June, the plants at the lower altitudes usually displaying the earliest blossoms. I have recorded the first flowers as early as April 11 (1945), but there is a specimen with flowers in the park herbarium that Fleetwood collected along Abrams Creek on March 31 (1935). As late as July 27 (1961) I noted a few blossoms still persisting on plants growing near the summit of the Chimney Tops. In 1945 the period of flowering in the park spanned three months. In the

Buckhorn area near Emerts Cove a shrub I had under observation held some blossoms from May 1 through May 31 (1955).

The largest recorded specimen of mountain-laurel [Dixon, 1961] grows at an altitude of approximately 4000 ft. on the east slope of Greenbrier Ridge, about 1 mile north of Greenbrier Knob. It towers to a height of about 25 ft., and the largest of its numerous trunks—fused at the base to form an aggregation measuring 20 ft. 6 inches in circumference—is 3 ft. 6 inches around. In 1951 this specimen was still in good condition, but eight years later I discovered much damage from what I believe to be the weight of heavy snows. Many large tree-like specimens grow on the slopes of Mt. Cammerer, near Soco Gap, and elsewhere.

Pieris D. Don

FETTER-BUSH—*Pieris floribunda* (Pursh) B. & H.

This evergreen shrub is very uncommon in the park, growing mostly at middle altitudes in open oak and pine stands [Shanks, 1954a]. It is one of the species comprising Whittaker's xeric shrub union. Flowers have been noted from March 22 (1938) to June 4 (1942). The fetter-bush is endemic in the Southern Appalachian Mountains.

Lyonia Nutt.

MALEBERRY*—*Lyonia ligustrina* (L.) DC.

This shrub occurs mostly in closed oak forests and in open oak and pine stands [Shanks, 1954a]. It is one of the species in Whittaker's [1956] subxeric heath union. Specimens have been noted from near the lowest (1100 ft.) to near the highest (6500 ft.) altitudes. The maleberry occurs on a number of the grass balds (Parson, Gregory, Andrews, Soco) as well as on Mt. Le Conte (Myrtle Point), Greenbrier Pinnacle, High Rocks, Black Camp Gap, Tuckaleechee Cove, Cades Cove, Happy Valley, and elsewhere. The flowering season extends from late May to middle July.

* The variety *foliosiflora* (Michx.) Fern. occurs on Andrews Bald ar possibly elsewhere [Jennison, 1939a; Sharp et al., 1960; Shanks, 1961].

Leucothoë D. Don

1. DOG-HOBBLE*—*Leucothoë fontanesiana* (Steud.) Sleumer

This evergreen shrub is of common occurrence in the park, "sufficiently abundant in places to form almost impenetrable thickets" [Jennison, 1939a]. It grows in the shade in moist acid soils. Shanks [1954a] listed it as a shrub of the cove hardwood, hemlock, and northern hardwood forests. Whittaker's [1956] "mesic heath union" consists of this plant and the rosebay rhododendron. The dog-hobble ranges in altitude from 1100 ft. (Abrams Creek) to 5800 ft. (Mt. Le Conte and Andrews Bald).

This shrub is in flower, ordinarily, from late April to early June, the extreme dates being March 30 (1938) and June 27 (1940).

2. REDTWIG LEUCOTHOË—*Leucothoë recurva* (Buckley) Gray

This deciduous shrub is rather scarce in the park occurring in "well-drained acid soils on ridges and slopes" [Jennison, 1939a]. It has been reported from along the Appalachian Trail between Inadu and Cosby Knobs (5100 ft.), Mt. Sterling (2700-3700 ft.), Greenbrier Pinnacle (4200 ft.), and Mt. Cammerer. Flowers were noted from April 23 (1934) to May 22 (1936). The leaves turn bright red in color in October.

Both species of *Leucothoë* that occur in the park are Southern Appalachian endemics—the evergreen *L. fontanesiana* and the deciduous *L. recurva*.

SOURWOOD—*Oxydendrum* DC.

SOURWOOD—*Oxydendrum arboreum* (L.) DC.

This is an abundant small tree at low and middle altitudes, especially in closed oak forests and in open oak and pine stands [Shanks, 1954a]. It is one of the species comprising Whittaker's [1956] subxeric tree class. The sourwood ordinarily reaches its

* Until recently this plant was known either as *L. catesbaei* or *L. editorum.*

uppermost altitude between 4500 and 5000 ft., with the highest reported elevation being 5600 ft. on Silers Bald [Gilbert, 1954].

The attractive panicles of fragrant white flowers serve to decorate this tree from late June throughout July, at which time they attract large numbers of bees. These flowers are the source of an excellent honey. The extreme dates of flowering are June 20 (1942, 1951) and August 22 (1946). The seed capsules, resembling the flowers in size and arrangement, persist into the winter season or later. The leaves turn scarlet in late summer and in the autumn; they account for much of the brilliant red coloration in second-growth and oak-dominated forests.

On a ridge between Lowes and Cannon Creeks in the Greenbrier area, at an altitude of 2600 ft., grows a record-size specimen of sourwood measuring 7 ft. 7 inches in circumference, approximately 75 ft. high, and with a spread of about 40 ft.*

TRAILING ARBUTUS—*Epigaea* L.

TRAILING ARBUTUS—*Epigaea repens* L.

This is a fairly common evergreen shrub especially on rather dry mountainsides where pine and oak forests prevail. Shanks [1954a] lists it as a ground-cover species in closed oak forests and a ground heath in open oak and pine stands. Along some trails such as on Hannah Mountain and the Abrams Falls Trail from Cades Cove it is quite common. Above 5000 ft. it is rather scarce, but I have noted it up to 5800 ft. on Andrews Bald.

Since its altitudinal range covers approximately 5000 ft., the flowering of this plant may extend over many weeks. In 1947 I noted the earliest blossoms at a low elevation on January 26; the species was in flower on Andrews Bald (5800 ft.) four months later. The latest blossoms were noted on Andrews Bald on May 29 (1939). Ordinarily, March and early April are the times of peak flowering at low and middle altitudes.

* The record given by the American Forestry Association [Dixon, 1961] is 7 ft. 4 inches in circumference—a tree growing in the Pisgah National Forest, N. C.

WINTERGREEN—*Gaultheria* L.

WINTERGREEN—*Gaultheria procumbens* L.

This "nearly herbaceous evergreen shrub with slender creeping stems and aromatic leaves" [Sharp, 1942b] is of common occurrence in acid soils at low and middle altitudes. Shanks [1954a] lists it as a ground heath of the open oak and pine stands. According to Sharp [1942b], its uppermost range is 5000 ft. Flowers have been noted over a two-months' period, beginning in middle June.

The wintergreen is often found associated with trailing arbutus *(Epigaea repens).*

HUCKLEBERRY—*Gaylussacia* HBK.

1. BLACK HUCKLEBERRY—*Gaylussacia baccata* (Wang.) K. Koch

This low shrub of the open oak and pine stands [Shanks, 1954a] is fairly common, especially at low and middle altitudes in the western half of the park. Whittaker [1956] includes it as a member of the xeric shrub union. The black huckleberry has been noted from 1800 ft. (Cades Cove) to approximately 5000 ft. (Gregory Bald). Flowers have been observed from early May to early June. The black fruits are sweet when ripe. "This species has glands on both surfaces of its leaves and this character, even in the vegetative condition, serves admirably to separate it from *G. ursina* (Curtis) T. & G. which has glands only on the lower surface of its leaves."*

2. BUCKBERRY—*Gaylussacia ursina* (M. A. Curtis) T. & G.

This shrub, also called "bear huckleberry," is of common occurrence ranging from 1800 ft. (Cades Cove) to 6600 ft. Whittaker remarked that it is "most important in oak-chestnut forests and under *Kalmia* in oak-chestnut heath ranging widely into other types but never to the mesic or xeric extremes."† Shanks [1954a] listed it as a shrub of the cove hardwoods and of the closed oak

---

* Letter of November 22, 1937, from W. H. Camp to F. H. Miller.
† Letter of February 3, 1949, from R. H. Whittaker to A. Stupka.

forests. Specimens have been known to attain a height of up to 15 ft. Flowers have been noted from late April and on throughout May. The fruits are ripe in August and early September.

The buckberry is one of the endemic plants of the Southern Appalachian Mountains.

BLUEBERRY—*Vaccinium* L.

Jennison [1939a] wrote that "Some of the true blueberries . . . found here are so imperfectly understood as to make it difficult, if not impossible to refer them correctly to known categories. . . . Before the genus *(Vaccinium)* can be satisfactorily treated, extensive collections and critical studies will have to be made." Hybridization is of frequent occurrence, making classification quite confusing. The species mentioned here are based in part on Camp [1945, and earlier correspondence], Whittaker [1956, and earlier correspondence], Shanks [1947], Sharp [1942b], Jennison [1939a], and others.

1. TREE SPARKLEBERRY—*Vaccinium arboreum* Marsh.

This rather rare shrub appears to be restricted to dry soils at low elevations. It is known to occur near Cades Cove (Cobb Ridge above Mill Creek), along lower Abrams Creek, and near the park at Laurel Lake. Its uppermost altitudinal range is at 2600 ft. on Cobb Ridge where a specimen was noted in full flower on May 26 (1949). Other flowering dates range from middle May to early June.

The small oval-shaped leaves of this distinctive blueberry are glossy and leathery, making identification relatively easy. Occasionally it grows large enough to be classed as a small tree. The black berries are inedible.

2. CONSTABLE'S BLUEBERRY*—*Vaccinium constablaei* Gray

Camp [1945] writes that this highbush blueberry occurs in "Western North Carolina and eastern Tennessee;† limited to the mountain tops and upper slopes, rarely below 3500 ft. elev., usually a prominent component of the ericaceous meadows . . . characteristic of the higher regions, where it associates with *Rhododendron, Kalmia,* and various other genera of *Ericaceae.*" Gilbert [1954], who lists this tall-growing shrub from Spence Field, Silers, Andrews, Gregory, and Parson Balds, calls it the most common and typical shrub on these high mountain meadows. Whittaker [1956] places Constable's blueberry in his subxeric heath union, along with *Kalmia latifolia, Lyonia ligustrina,* and *Smilax glauca.*

Camp [1945] regards the fruit as "often of excellent flavor." He mentions a plant of this species which he and the late H. M. Jennison found in 1936 "near the old Bote Mountain trail" that was approximately 26 ft. in height. "So far as is known, this is the tallest *Vaccinium* plant in eastern North America" [ibid.].

3. MOUNTAIN-CRANBERRY‡—*Vaccinium erythrocarpum* Michx.

This Southern Appalachian endemic is common throughout the spruce-fir zone and occurs, in places, down to 3500 ft. Oosting and Billings [1951] call it one of the three species of shrubs "present in almost every stand" of the Southern Appalachian spruce-fir forests that they studied (the other two: *Rubus canadensis* and *Sambucus pubens).* Whittaker includes it with his "low-shrub union of subalpine forests." It occurs at or near the

---

* According to Camp [1945], Constable's blueberry was derived from combinations between the upland highbush blueberry *(V. simulatum)* and the mountain dryland blueberry *(V. alto-montanum).* "As in other complexes of this type, local populations have sometimes been built up around one set of characters, while in others somewhat different phases may be dominant. For example: the 'simulatoid' phase is common along the crest of the mountains and especially the contiguous upper slopes in the region of Gregory Bald and Thunderhead in the Great Smoky Mountains. On Mt. Le Conte, the 'alto-montanoid' phase dominates on some of the exposed and storm-swept ridges, whereas the 'simulatoid' phase usually is the only one found in the tangled ericaceous 'slicks' " [ibid.].

† Gleason [1952] adds Virginia, West Virginia, and Kentucky to its range.

‡ According to Sharp et al. [1960] and Shanks [1961], the form *nigrum* Allard is represented in the park.

summits of some of our highest mountains (6500 ft.).

June is the month of most flowering, the dates ranging from May 25 (1935) to July 19 (1936). The corolla is four-cleft, and its deeply-cut lobes are strongly recurved. The fruit ranges in flavor from rather insipid to sweet.

4. HAIRY BLUEBERRY—*Vaccinium hirsutum* Buckl.

This low-bush blueberry is rather uncommon, and it appears to be restricted to that part of the park from Thunderhead westward— essentially the westernmost quarter of the area. It is a Southern Appalachian endemic whose range is given as "Great Smoky Mts. and outliers" by Camp [1945]. In altitude it extends from 1700 to approximately 5000 ft. where it occurs "in sandy loams on wooded slopes and summits" [Jennison, 1939a]. Shanks [1954a] lists it as one of the lower shrubs of the open oak and pine stands, while Whittaker [1956] considers it as a member of his "xeric shrub union with low Vaccinioideae dominant."

Flowering dates range from April 28 (1948, Cades Cove) to July 2 (1934, Gregory Bald). The hairiness of the fruits serves to distinguish this plant from all the other blueberries.

Localities where specimens occur include Spence Field Bald, Gregory Bald, Parson Bald, Rich Mountain, Cades Cove, Crib Gap, Panther Gap, Scott Gap, and Chilhowee Mountain.

5. UPLAND LOW BLUEBERRY*—*Vaccinium pallidum* Ait.

It was Jennison's [1939a] belief that this was "probably the most common and widely distributed blueberry in our area." Specimens in the local collections would indicate that its range is from 2500 to 5800 ft., although Camp [1945] states that it is found "generally below 3500 ft. elev.; under 2500 ft. elev., often grading into one of the phases of *V. vacillans*." This is one of the low-bush blueberries whose usual habitat, according to Sharp [1942b], is moist acid soils.

* Camp [1945] who proposed the name "Upland low blueberry" writes that "It is admittedly difficult to know just where to draw the line between the upland *pallidum* and the lowland *vacillans,* for they have hybridized in the past and the segregate forms today fill the gap, both morphologically and distributionally, between the two basic populations. . . . But in spite of the presence of many confusing individuals, a considerable amount of reasonably pure material of both *pallidum* and *vacillans* still exists." Whittaker [1956] calls the mixture *V. vacillans-pallidum.*

Records of flowering extend from mid-April to mid-June. Ripe fruit has been noted from early July to middle August.

6. UPLAND HIGHBUSH BLUEBERRY*—
Vaccinium simulatum Small

Shanks [1954a] lists this as a shrub of spruce-fir and closed oak forests, and as a tall shrub in open oak and pine stands. According to the local herbarium specimens, its altitudinal range is from 2500 to 5500 ft., and its time of flowering extends from mid-April to mid-June. Ripe fruits have been noted from early July to early September. Jennison [1939a] indicated that its distribution and habitats are as for *V. pallidum.*

7. DEERBERRY†—*Vaccinium stamineum* L.

This common shrub is usually to be found growing in open oak and pine stands [Shanks, 1954a]. According to Jennison [1939a], it occurs in well-drained sandy acid soil on sparsely wooded slopes and ridges. Whittaker [1956] includes it as a member of his sub-xeric shrub union. Its altitudinal range is from 1500 to 5000 ft. It is to be found on Spence Field, Gregory, Parson, and Soco Balds.

The deerberry, locally called "gooseberry," has numerous clusters of white bell-shaped flowers which make it a conspicuous shrub when in bloom. Flowers have been noted from April 11 (1945) at a low altitude to June 28 (1947) on Gregory Bald.

The large round fruits drop to the ground soon after ripening. In color they vary from greenish to yellowish and pale purple. From the thick-skinned, juicy, sour-tasting fruits a jam of high quality is made by some of the local people.

8. LOW BLUEBERRY‡—*Vaccinium vacillans* Torr.

Shanks [1954a] lists this as a low shrub of the open oak and pine stands. Whittaker [1956], who prefers to group the upland low

* Jennison [1939a] regarded *simulatum* as a variety of *V. pallidum* "from which it is distinguished with difficulty." The name "Upland highbush blueberry" was proposed by Camp [1945].

† According to Shanks [1961], "our variable populations include var. *candicans* Mohr, var. *melanocarpum* Mohr, and var. *neglectum* (Small) Deam."

‡ Refer to footnote under *V. pallidum.*

blueberry *(V. pallidum)* with this species since the two have a tendency to blend together in some areas, calls it a member of his xeric shrub union. Ordinarily less than 2 feet high, this common shrub grows in well-drained sandy soils [Sharp, 1942b] from approximately 1500 to 5000 ft.

.   .   .   .

In addition to the foregoing species of blueberries, the following have appeared either on some of the lists of park plants [Jennison, 1939a; Sharp, 1942b; Shanks, 1961] or in published articles [Camp, 1945].

MOUNTAIN DRYLAND BLUEBERRY—
Vaccinium alto-montanum Ashe

This shrub is listed for Great Smoky Mountains National Park by Shanks [1961], but since Camp [1945] calls attention to the similarity between it and Constable's blueberry and since there appear to be no specimens in the University of Tennessee nor in the park herbarium, it is a plant whose status here remains in doubt. (Also see first footnote under *Vaccinium constablaei.)*

BLACK HIGHBUSH BLUEBERRY—
Vaccinium atrococcum (Gray) Heller

A number of the local herbarium specimens both at the University of Tennessee and in the Great Smoky Mountains National Park bear this name, but this species is not listed for the park by Sharp et al. [1960] nor by Shanks [1961], and in 1937 Camp, in his correspondence with F. H. Miller, advised him to omit it from a list of park plants. According to Camp [1945], *V. atrococcum* has been confused with *V. corymbosum.* (In the 1857 revised edition of *Gray's Manual, atrococcum* was given as a variety of *V. corymbosum.)*

NORTHERN HIGHBUSH BLUEBERRY*—
Vaccinium corymbosum L.

Some of the blueberry plants collected in the park by Jennison and others in the 1930's were identified as being this species, but Camp, in a letter to F. H. Miller in 1937, suggested that it be

* The name "Northern highbush blueberry" was proposed by Camp [1945].

deleted from a list of park plants. According to both Camp [1945] and Gleason [1952], this shrub would be out of range here.

LAMARCK'S SUGAR BLUEBERRY—
Vaccinium lamarckii Camp

Camp [1945] gives the range of this blueberry as "Newfoundland, west to Minnesota, southward usually in the uplands and on mountains to West Virginia and North Carolina (the Great Smoky Mts.)." Although there are no specimens with this determination in the herbaria in Great Smoky Mountains National Park and the University of Tennessee, a plant originally called *V. angustifolium* Ait., which Sharp and F. H. Miller collected on Horse Ridge in June 1942, may be referable to this species. Gleason [1952] calls attention to the close resemblance between *V. lamarckii* and *V. angustifolium*. Fernald [1950] refers *lamarckii* to *V. angustifolium laevifolium* whose southern limits he gives as Virginia and West Virginia.

FAMILY ✦ *EBENACEAE*

PERSIMMON—*Diospyros* L.

COMMON PERSIMMON—*Diospyros virginiana* L.

This is a fairly common tree at altitudes up to about 2500 ft. It occurs most frequently along roadsides and in old fields "in well drained sandy soils" [Sharp, 1942a]. A number of these trees grow in Cades Cove and vicinity.

Late May and early June marks the time of its peak of flowering; the range of blossoming has been from May 18 (1948) to June 20 (1942). The fruits become sweet and quite edible after early frosts. Bears, opossums, skunks, gray foxes, and other animals feed on the ripe persimmons.

FAMILY ✦ *SYMPLOCACEAE*

SWEETLEAF—*Symplocos* Jacq.

COMMON SWEETLEAF—*Symplocos tinctoria* (L.) L'Her.

There is but one record of this rare shrub from the park. On May 4, 1937, Jennison [1939a] collected a flowering specimen at Epps Spring on Canebrake Creek near Bryson City "in sandy alluvium along the Tuckasegee River." That site was inundated by Fontana Reservoir in the middle 1940's when Fontana Dam was completed, and no additional specimens of the common sweetleaf have been located in that vicinity or elsewhere in the park.

FAMILY ✦ *STYRACACEAE*

SILVERBELL—*Halesia* Ellis

MOUNTAIN SILVERBELL—*Halesia carolina* var. *monticola* Rehd.

Jennison [1939a] regarded the silverbell as "one of the most frequent trees in our area." Shanks [1954a] listed it as a principal tree of the cove hardwoods, northern hardwoods, hemlock forests, and closed oak forests. According to Ayres and Ashe [1905], silverbell (or "peawood" as it was called) made up about 10 percent of the forest cover of the lower Little River basin at the turn of the century. Whittaker [1956], who places it in his mesic tree class, indicates that it extends up the slopes to 5700 ft., but the species is rather scarce above 5000 ft.

Ordinarily the silverbell is in full bloom from middle April to middle May, but the flowering season may span a period of two months, as it did in 1945 (March 21-May 24). Extreme dates of flowering are March 21 (1945) and June 6 (1947). Seeds in the distinctive four-winged fruits are eaten by squirrels in late summer and in the autumn. On silverbells near Newfound Gap the fruits have been known to remain as late as January.

A tree on False Gap Prong and another on Cannon Creek measure 11 ft. 9 inches in circumference; according to the American Forestry Association, these are the largest silverbells on record [Dixon, 1961]. Along Surry Fork there are at least five trees that measure over 9 ft. in circumference.

FAMILY ✦ *OLEACEAE*

ASH—*Fraxinus* L.

1. WHITE ASH*—*Fraxinus americana* L.

This is a fairly common tree which, according to Shanks [1954a], is one of the dominant species in cove hardwood and in hemlock forests. It grows mainly in rich moist woodlands where it ranges from the lowest altitudes to 5200 ft. [Whittaker, 1956]. Small specimens grow on Spence Field Bald and Gregory Bald near 5000 ft. At the turn of the century white ash comprised about 5 percent of the forest cover in the Laurel Creek basin according to Ayres and Ashe [1905].

Flowers may appear in late March, and some trees are in flower in late April and possibly later at higher altitudes.

Specimens up to 15 ft. 5 inches in circumference grow in the park, this being the measurement of one on Kalanu Prong (Greenbrier area).† One measuring 13 ft. 1 inch in circumference grows along a trail below the Chimneys Parking Area; another, 12 ft. 6 inches in circumference, is identified along the course of the Big Locust Nature Trail (Chimneys Campground).

2. GREEN ASH‡—*Fraxinus pennsylvanica* Marsh.

This tree is uncommon in the park where it grows in moist soils at low altitudes. Low woodlands and streamsides are a favored habitat. A specimen collected by Whittaker near Porters Flat

---

* According to Little [1953], this includes the so-called Biltmore ash (*F. biltmoreana*) which, in the past, was considered either as a full species or as a variety of *F. americana*.

† The largest reported to the American Forestry Association is a tree 22 ft. 3 inches in circumference, in Pennsylvania [Dixon, 1961].

‡ According to Little [1953], this includes the variety *subintegerrima* (Vahl) Fern. listed by Shanks [1961] and the variety *lanceolata* (Borkh.) Sarg. listed by Sharp [1942a].

at 3000 ft. appears to represent its uppermost limit. It has been noted in Cades Cove, Gatlinburg, and along lower Abrams Creek.

LILAC—*Syringa* L.

COMMON LILAC—*Syringa vulgaris* L.

This cultivated shrub was introduced about some old homesites where a few plants may still persist.

Forsythia Vahl

Forsythia spp.

Some of these exotic shrubs still persist around old homesites where they were introduced many years ago.

FRINGETREE—*Chionanthus* L.

FRINGETREE—*Chionanthus virginicus* L.

This is a very rare shrub in the park, the only specimen from within the area coming from lower Abrams Creek at about 900 ft. altitude. It occurs close to the park in Wears Cove and at Henderson Springs. In her report on the trees and shrubs of East Tennessee, Galyon [1928b] remarks that the fringetree grows along streams, is "Commonly cultivated; sometimes escaping."

Flowers have been noted from April 16 (1945) to May 8 (1937).

PRIVET—*Ligustrum* L.

EUROPEAN PRIVET—*Ligustrum vulgare* L.

This exotic shrub was introduced about a number of old homesites where it persists.

FAMILY ✦ *APOCYNACEAE*

PERIWINKLE—*Vinca* L.

COMMON PERIWINKLE—*Vinca minor* L.

This exotic trailing shrub was "often planted in cemeteries and gardens" [Jennison, 1939a] where the plants have persisted.

FAMILY ✦ *VERBENACEAE*

Callicarpa L.

FRENCH MULBERRY—*Callicarpa americana* L.

This is a rare shrub in the park, having been recorded only from Cades Cove and from Little River (about a mile below the Sinks and near the confluence with the Middle Prong). It grows "in shaded gravelly soil, banks of water courses, at lower elevations" [Jennison, 1939a]. The French mulberry grows near the park in Tuckaleechee Cove and Wears Cove where flowers have been noted in July.

"The rough opposite leaves and thickly clustered fruits, magenta when ripe, characterize this species" [Sharp, 1942b].

FAMILY ✦ *BIGNONIACEAE*

Paulownia Sieb. & Zucc.

ROYAL PAULOWNIA; PRINCESS-TREE—*Paulownia tomentosa* (Thunb.) Sieb. & Zucc.

This oriental tree is uncommon at low altitudes in the park. It occurs along Little River, Laurel Creek, Big Creek, and along the western boundary. The royal paulownia grows abundantly outside the park in the Nantahala Gorge southwest of Bryson City, North Carolina; it was also introduced in many nearby communities. April is the usual month of bloom, although the large attractive clusters of lavender-blue flowers have

Contrasting with the dark evergreen spruce forest is a beech gap along the main crest of the Great Smoky Mountains near Indian Gap. The ability of American beech *(Fagus grandifolia)* to withstand considerable wind damage is regarded as the major factor in its continued survival in such locations.

Coniferous forest at approximately 6000 ft. elevation along the road to Clingmans Dome. Fraser firs *(Abies fraseri)* make up more than 80 percent of the stand, the remainder being red spruces *(Picea rubens)*.

Evergreen plants of the high altitudes—Fraser fir *(Abies fraseri)* above, Huger's sand myrtle *(Leiophyllum buxifolium hugeri)* center, and piedmont rhododendron *(Rhododendron minus)* at lower left.

This yellow birch *(Betula alleghaniensis)* germinated on the top of a decaying stump. When the stump disappears the tree will be on stilts or prop-roots.

Outlining the ridges in the region of Huggins Hell—the wild country on the south slope of Mt. Le Conte—are dense evergreen growths locally called "laurel slicks." These are practically impenetrable tangles comprised mostly of Catawba rhododendron *(Rhododendron catawbiense)* and mountain laurel *(Kalmia latifolia).* Red spruce *(Picea rubens)* is the dominant forest tree.

Following a heavy snowfall—forest trees as viewed from a ledge near Ramsey Cascades. At this altitude (4200 ft.) spruces *(Picea rubens)* are approaching their lowest limits in these mountains.

been noted from March 23 (1938) to May 13 (1941). (The name commemorates Anna Paulowna, 1795-1865, Queen of the Netherlands.)

Except for the color of the flowers and the appearance of the seed pods, this exotic tree resembles the catalpa. The leaves are large, particularly on some of the sprouts; an abnormally large leaf growing on a sprout near the park boundary close to Davenport Gap measured 37 inches wide and 25 inches long (August 10, 1954).

TRUMPET-FLOWER—*Campsis* Lour.

TRUMPET-CREEPER—*Campsis radicans* (L.) Seem.

This woody vine is rather uncommon, growing mostly along or near the park boundary at low altitudes. Its occurrence in the Sugarlands near 2000 ft. altitude may indicate introduction by persons living there in pre-park days. It has also been noted in Cades Cove, along the lower part of Little River, and along lower Abrams Creek. Near the park it is not uncommon along roads and fencerows, especially in the vicinity of the western boundary.

The orange-red trumpets have been noted from late May to middle July, with occasional flowers in August and September. This attractive plant is known locally as "cow-itch."

Bignonia L.

CROSS-VINE—*Bignonia capreolata* L.

This woody vine is fairly common at altitudes below 2000 ft. where it often grows along streams or in moist woods. The dark leaves are almost evergreen, persisting through the winter. At the time of its flowering—usually from late April to late May (earliest: April 12, 1963)—this vine ordinarily displays a generous quota of large orange- and yellow-colored blossoms, making it one of our most handsome native plants.

A cross-section of the stem reveals the wood-bundles in the form of a cross, hence the name.

Catalpa Scop.

NORTHERN CATALPA—*Catalpa speciosa* Warder

This tree grows in Cades Cove where, in all probability, it was introduced many years ago. The northern catalpa occurs naturally in West Tennessee and westward [Munns, 1938; Braun, 1961]. It likewise grows on lands and in communities near the park where it has been cultivated. Its flowering period extends from late May throughout June.

FAMILY ✦ *RUBIACEAE*

BUTTONBUSH—*Cephalanthus* L.

COMMON BUTTONBUSH—*Cephalanthus occidentalis* L.

This shrub is rare in the park where it grows only in a wet meadow and in a sphagnum-bordered marsh in Cades Cove at about 1800 ft. elevation. Flowers have been noted from late June throughout July.

PARTRIDGE-BERRY—*Mitchella* L.

PARTRIDGE-BERRY—*Mitchella repens* L.

This is a small semi-herbaceous trailing plant whose rounded evergreen leaves we associate with hemlock forests—where it appears to have its best growth. It is of common occurrence in the park from the lowest altitudes to approximately 5200 ft. This low creeping plant thrives in rich acid soils where the shade is heavy. The paired white four-lobed tubular flowers have been noted from May 12 (1938) to August 6 (1961); ordinarily the peak of bloom is in June. A single long-persistent red fruit results from the paired flowers; this fruit is rather insipid but edible.

On June 25, 1946, I noted a number of fragrant flowers of the partridge-berry with three- and five-parted corollas; the location was near 4000 ft. elevation along the trail from Cherokee Orchard to Trillium Gap.

FAMILY ✦ *CAPRIFOLIACEAE*

BUSH-HONEYSUCKLE—*Diervilla* Duham.

BUSH-HONEYSUCKLE—*Diervilla sessilifolia* Buckl.

This is a common shrub of the spruce-fir forests [Shanks, 1954a], occurring but rarely below 4000 ft. Whittaker [1956] includes it with the "low-shrub union of subalpine forests." On some of the higher mountains the bush-honeysuckle extends to at least 6500 ft. According to Gilbert [1954], it is a prominent shrub on Mt. Sterling Bald; it also occurs on Andrews Bald and Silers Bald.

July is the usual month of flowering; the span of blossoming has been recorded from June 19 (1936) to August 14 (1942). The yellow flowers are scentless.

The bush-honeysuckle is one of the endemic plants of the Southern Appalachian Mountains [Harper, 1947].

HONEYSUCKLE—*Lonicera* L.

1. FLY HONEYSUCKLE—*Lonicera canadensis* Bartr.

This straggling shrub is localized and quite scarce in the park where it is at, or near, its southernmost limit. It grows at altitudes ranging from 3000 to 6000 ft. in moist rocky situations. Most of the records of its occurrence are from the north and northwest slopes of Mt. Le Conte (Roaring Fork and Le Conte Creek) at altitudes above 4000 ft.; it has also been found near 6000 ft. along the trail between Andrews Bald and the Clingmans Dome Parking Area where it is rare. I have noted it near the West Prong of the Little Pigeon River at approximately 3000 ft. (above Chimneys Campground) which appears to be the lowest altitude at which it grows.

The paired yellowish-green flowers have been observed as early as March 31 (1937, at the 3000 ft. elevation) and as late as May 17 (1942, at a high altitude).

2. JAPANESE HONEYSUCKLE—*Lonicera japonica* Thunb.

In pre-park days this Asiatic vine was introduced at many a homesite and cemetery where it persists and spreads forming a dense ground- and plant-cover. The aggressiveness of this pernicious weed is demonstrated best on lower Hazel Creek where it has run rampant over the ruins of the extinct community of Proctor, now a part of the park. Flowering is mostly in May and June.

3. TRUMPET HONEYSUCKLE—*Lonicera sempervirens* L.

There is some question as to whether this very uncommon vine should be regarded as one that has escaped from cultivation. Both Jennison [1939a] and Sharp [1942b] were of the belief that the trumpet honeysuckle is not indigenous, since its occurrence is usually near former homesites. Specimens have been found near Cherokee Orchard and near the park line in Emerts Cove. Flowering takes place in May and June.

Symphoricarpos Duham.

CORALBERRY—*Symphoricarpos orbiculatus* Moench

In view of the localities where this shrub grows—along roadways and near old homesites—it is, in all probability, a species that has become naturalized. The coralberry with its dense clusters of small persistent fruits is frequently planted and escapes from cultivation. It is rather scarce in the park, occurring in the Ravensford, Greenbrier, Sugarlands, Elkmont, and Cades Cove areas. Flowers have been noted in early July.

Viburnum L.

1. MAPLE-LEAVED VIBURNUM*—*Viburnum acerifolium* L.

This is a fairly common shrub "in well drained sandy loam on woody slopes and ridges" [Jennison, 1939a]. It is listed as a shrub of the cove hardwoods by Shanks [1954a] and as one of the species in Whittaker's [1956] submesic shrub union. It ranges from the lowest elevations to 4700 ft. (Ledge Bald). May and early June are the usual months of flowering; blossoms have been noted from April 28 (1948, Wears Cove) to June 17 (1961, Hannah Mountain).

This is a particularly attractive shrub in the fall of the year when its maple-like foliage becomes pinkish, purplish, or magenta in color.

* According to Shanks [1961], "the common form is var. *glabrescens* Rehd." There is a specimen of the variety *ovatum* McAtee from near the summit of Gregory Bald in the University of Tennessee Herbarium.

2. HOBBLEBUSH—*Viburnum alnifolium* Marsh.

This straggling shrub grows from 3000 ft. to near the summits of some of our highest mountains; above 4500 ft. it is quite plentiful in cool damp woodlands. Shanks [1954a] listed it as a shrub of the cove hardwood, hemlock, northern hardwood, and spruce-fir forests.

The latter half of April and the month of May mark the peak of flowering; extreme dates are March 31 (1945) and June 12 (1940). The large round leaves may begin to color by late July, in places, and the change from green to varigated hues continues throughout August and into September. In the latter half of September the hobblebush is at the peak of its autumnal leaf color and is then one of the most strikingly-arrayed shrubs in the park.

3. WITHEROD—*Viburnum cassinoides* L.

Although this shrub has been recorded from the lower altitudes (mouth of Abrams Creek; Whiteoak Sink), it is quite scarce below 4000 ft.; above 4000 ft. the witherod is fairly common. Shanks [1954a] lists it as a shrub of the spruce-fir forest, while Whittaker [1956] includes it as a member of his "lower-elevation heath bald shrub union."

Flowering is at its height in June and early July, the extreme dates being May 31 (1936) and July 17 (1940).

4. ARROW-WOOD*—*Viburnum dentatum* L. (combination)

Specimens of this very uncommon shrub have been noted ranging from 900 ft. (near mouth of Abrams Creek) to 4650 ft. at Moores Spring (near Gregory Bald). At the latter place I have recorded flowers from middle to late June.

* Neither Jennison [1939a] nor Sharp et al. [1960] include this species in their lists of park plants, but it is included by Shanks [1961]. McAtee [1956] calls attention to the highly variable *dentatum* complex "with the characters intergrading in apparently all directions." Some of these variants have been proposed as representing new species. Since the following have been associated with *V. dentatum* in one way or another, I am including them with that species: *V. carolinianum* Ashe, *V. recognitum* Fern., *V. scabrellum* T. & G., and *V. semitomentosum* Michx.

5. SNOWBALL*—*Viburnum opulus* L.

A few specimens of this European shrub continue to grow near old home-sites where they were introduced in pre-park days (Greenbrier; Smokemont).

6. BLACKHAW—*Viburnum prunifolium* L.

This rare shrub is known only from the Cades Cove area of the park. It also occurs in the vicinity of the park in Tuckaleechee Cove, Pigeon Forge, and near Gatlinburg. Flowers have been noted from March 31 (1945, an exceptionally warm month) to April 28 (1943). This plant becomes a small tree on occasions. Its smooth leaves resemble the leaves of some plums.

7. RUSTY BLACKHAW—*Viburnum rufidulum* Raf.

Like the blackhaw (*V. prunifolium*) which it resembles rather closely, the rusty blackhaw is rare in the park having been found only in Whiteoak Sink. It grows in limestone soils in Wears Cove, close to the park, where it becomes a small tree. Flowers have been noted there in late April and early May. The buds and the undersurfaces of the thick leaves and leaf stems are covered with rust-colored hairs.

ELDER—*Sambucus* L.

1. AMERICAN ELDER—*Sambucus canadensis* L.

This is a common shrub growing in "open woods, clearings, river-banks, and swampy places" [Jennison, 1939a] from the lowest altitudes to approximately 5200 ft. The American elder occurs at Soco Gap (approx. 4300 ft.), Pin Oak Gap (4428 ft.), Black Camp Gap (4522 ft.), and at approximately 5150 ft. between Newfound and Indian Gaps. Jennison, in a note pertaining to a specimen collected at 5500 ft. on the Appalachian Trail near Spence Field, believed it may have been introduced at that high altitude by stock and stockmen.

Ordinarily the flowering period begins in late May and continues throughout June. Along the Little Tennessee River near

---

* The variety *roseum* L. is the most familiar—a large round head of showy blossoms whose flowers are sterile and enlarged.

the mouth of Abrams Creek the American elder has come into flower as early as May 18 (1948), while at high altitudes (vicinity of Newfound Gap) blossoms have been noted as late as August 26 (1953). The fruits ripen in August and September.

2. RED-BERRIED ELDER—*Sambucus pubens* Michx.

This is one of the few woody plants whose range extends from near the lowest to near the highest elevations in the park. Although it is rather scarce in the lower altitudes, it is of common occurrence above 4000 ft. Oosting and Billings [1951] listed it as one of the species present in almost every stand of spruce-fir forest they investigated in the Southern Appalachian Mountains. According to Shanks [1954a], it is a shrub of the hemlock and spruce-fir forests.

Ordinarily the red-berried elder is in flower from late April to middle June, but the extreme dates have ranged from March 17 (1938) to June 28 (1955). In 1938 the span of the flowering of this species covered a period of practically 3 months—from March 17 to June 15. At the lower altitudes the fruits may become bright red in color as early as late May, but in the spruce-fir regions where this shrub is much more plentiful the month of July marks the peak of fruit color.

Specimens of this elder have been known to reach a trunk diameter of 4 inches and a height and spread of 10 feet.

Glossary

ACUMINATE. Tapering to a slender point

ACUTE. Forming a sharp angle

ALTERNATE. Not opposite to each other; leaves borne one at a node at regular intervals

AMENT. A catkin, its numerous component flowers being without petals

APPRESSED. Pressed close to, or lying flat against

AURICULATE. With lobes or appendages at the base

AXILLARY. Relating to the axis or main stem

BILATERAL. Relating to two corresponding sides

BIPINNATE. Doubly or twice pinnate

BLADE. The expanded part of a leaf, petal, or sepal

BRACT. A specialized leaf

BRANCHLET. The final or farthest divisions of a branch

CALYX. The outer, usually green or leafy, part of a flower, contrasted with the inner showy portion (corolla)

CAPSULE. A dry fruit usually containing two or more seeds

CILIATE. With marginal hairs

COLLATERAL. Side by side

COMPOUND. Blade divided into leaflets

CORDATE. Valentine-shaped

COROLLA. The petals of a flower, collectively

CRENATE. Margins scalloped

CRENULATE. Margins finely scalloped

CUNEATE. Wedge-shaped

DECIDUOUS. Not persistent; falling off at maturity

DECOMPOUND. More than once compound

DENTATE. Toothed; the teeth directed outward

DRUPE. A fleshy fruit with a hard inner portion

ENTIRE. Without teeth or divisions

EVERGREEN. Not deciduous

FASTIGIATE. Erect and close together

FOLIACEOUS. Leaf-like

FUSIFORM. Spindle-shaped; tapering at each end

142

GLABROUS. Smooth; lacking pubescence

GLAND. A secreting organ

GLANDULAR. Containing or bearing glands

GLAUCOUS. Having the surface covered with a grayish, powdery bloom or coating

HERBACEOUS. Having the characters of a plant that has no persistent woody stem above ground

HUSK. The outer covering of various seeds or fruits

INFLORESCENCE. The flowering part of a plant

INTERNODE. The portion of a stem between two nodes

LANCEOLATE. Lance-shaped; broadest toward the base

LEAFLET. One of the divisions of a compound leaf

LEGUME. Fruit (pod) of the *Leguminosae*

LENTICEL. A small corky spot or dot on young bark

MEMBRANOUS. Thin; soft; pliable

NODE. Place on a stem which ordinarily bears a leaf or leaves

OBTUSE. Blunt

OPPOSITE. Leaves borne two at a node on opposite sides of the stem

OVATE. Egg-shaped; broadest near the base

PALMATE. Leaflets radiate from the end of the petiole

PERSISTENT. Remaining long attached

PETIOLE. A leaf-stalk

PINNA (pl. pinnae). One of the principal divisions of a pinnate or compoundly pinnate leaf

PINNATE. Leaves arranged in two rows along the rachis

PITH. The spongy center of a stem

POME. Fruit resembling an apple

PUBESCENT. Bearing hairs on the surface

PULVINUS (pl. pulvini). An enlargement of the petiole at the point of attachment to the stem

PUNCTATE. Dotted with minute spots

RACEME. A type of inflorescence in which the elongated axis bears flowers on short stems, the lowest opening first

RACHIS. The continuation of the leaf-stem as the axis of a pinnately compound leaf

RENIFORM. Kidney-shaped

SAMARA. A dry winged fruit usually one-seeded

SERRATE. Having sharp teeth pointing forward

SERRULATE. Finely serrate

SESSILE. Without a stalk

SIMPLE. A form in which the blade is not divided into leaflets

SINUATE. Having a wavy margin

SINUS. The space between two lobes

SPHEROIDAL. Almost a sphere

SPINOSE. Spine-like

STELLATE. Star-shaped

STIPULAR. Pertaining to a stipule

STIPULE. One of a pair of small appendages borne near the base of the leaf-stem

TENDRIL. A slender twining organ serving as a means of attachment

TOMENTOSE. Wooly; covered with densely matted hairs

TRIFOLIOLATE. Having three leaflets

TRIPINNATE. Thrice pinnate

TRUNCATE. Having the end square, as if cut off

VALVATE. Meeting only along the margins; not overlapping

Keys ————————————————————

to Trees, Shrubs, and Woody Vines of Great Smoky Mountains National Park *

By A. J. Sharp and Arthur Stupka

Keys to the species within the genera will be found in the alphabetical list of genera which follows each of the three main categories (Trees; Shrubs; Woody Vines). When a genus is represented in the park by only one woody species its full name is given in the key to the genera.

KEY TO TREES

1. Deciduous trees (leaves of one year falling before the subsequent leaves expand) 3
1. Evergreen trees (leaves persisting into the second year or longer; leaves broad, hard, leathery, or needle-like or scale-like) 2
   2. Trees with needle-like leaves or small scale-like leaves.......*Group A*
   2. Trees with broad, hard, leathery leaves...............................*Group B*
3. Thorn-bearing trees; branches and sometimes trunks bearing various kinds of thorns...*Group C*
3. Trees without thorns 4
   4. Trees with two or more leaves at a node............................*Group D*
   4. Trees with alternate leaves (leaves, leaf-scars, lateral branches, and buds characteristically one at a node)... 5
5. Leaves compound ....*Group E*
5. Leaves simple, sometimes deeply lobed but never with distinct leaflets ....*Group F*

* Grateful acknowledgment is made to Mrs. R. E. Shanks and the University of Tennessee Press for permission to use the 1960 *Summer Key to Tennessee Trees* in the preparation of the keys to trees. The original tree key is available from the University of Tennessee Press.

Group A

Key to the genera with needle- or scale-like leaves

1. Leaves needle-like, 2, 3, or 5 in a bundle with a sheath at the base, leaves 2 in. or more long...*Pinus*
1. Leaves not in bundles, linear or scale-like, less than 2 in. long........... 2
   2. Leaves all small, scale-like, overlapping, the leafy twig more or less flattened ....*Thuja*
   2. Leaves not all small and scale-like, not on flattened twigs.............. 3
3. Leaves produced more or less in one plane....................................... 6
3. Leaves spreading in all directions, at least on young growth........... 4
   4. Leaves flattened 5
   4. Leaves 4-angled, square in cross section; cones pendant..........*Picea*
5. Leaves awl-shaped, tapering to a sharp point, or scale-like.................
....*Juniperus virginiana*
5. Leaves linear; cones erect...*Abies*
   6. Leaves borne on stalks which perisist on the twigs after the leaves have fallen; cones pendant..*Tsuga canadensis*
   6. Leaf stalks not persisting on twigs; cones erect...............*Abies fraseri*

Group B

Key to the genera of broad-leaf evergreens

1. Leaf-margins undulate, all with a few stout spinose teeth.......*Ilex opaca*
1. Leaf-margins entire or with small teeth... 2
   2. Leaf-margins crenate-serrulate and ciliate..................*Pieris floribunda*
   2. Leaf-margins not ciliate 3
3. Leaves obovate to oblong, usually less than 1½ in. in length; fruit a berry ....*Vaccinium arboreum*
3. Leaves averaging more than 2 in.; fruit not a berry........................... 4
   4. Twigs encircled by a stipule scar at each node; fruit a cone............
....*Magnolia*
   4. Twigs not encircled by stipule scars... 5
5. Leaf-blades averaging more than 4 in. long......................*Rhododendron*
5. Leaf-blades averaging less than 4 in. long......................*Kalmia latifolia*

Group C

Key to the genera of trees with thorns on the stem

1. Leaves simple.. 5
1. Leaves compound.. 2
   2. Leaves at least in part decompound (more than once pinnate).. 4
   2. Leaves all pinnately compound (once pinnate).............................. 3
3. Leaflets ovate, rounded at both ends, not punctate....................*Robinia*
3. Leaflets ovate, pointed, glandular-punctate.....*Zanthoxylum americanum*
   4. Leaves very large (1½ ft. to 4 ft. long) bipinnate to tripinnate, borne in a cluster at the top of the stem; thorns simple.......*Aralia spinosa*

4. Leaves smaller (less than 1 ft. long), pinnate to bipinnate, scattered on the twigs; thorns stout, frequently branched *Gleditsia triacanthos*

5. Leaves entire ....*Maclura pomifera*

5. Leaves variously toothed, sometimes lobed..6

    6. Twigs typically stout, over ⅛ in. in diameter; the spur branches which are prolonged into thorns but little differentiated and often bearing leaves; leaves not deeply lobed...................................*Pyrus*

    6. Twigs typically slender, less than ⅛ in. in diameter; thorns differentiated, stiff and sharp; leaves often deeply lobed.................*Crataegus*

Group D

Key to the genera with 2 or more leaves at a node

1. Leaves simple...4

1. Leaves compound..2

    2. Leaves palmately compound (leaflets clustered at the apex of the petiole) ....*Aesculus octandra*

    2. Leaves pinnately compound or trifoliate.....................................3

3. Leaflets 3 to 5, coarsely toothed toward the apex; fruit double-winged ....*Acer negundo*

3. Leaflets commonly 7-11, entire or finely toothed; fruit single-winged.....
....*Fraxinus*

    4. Leaves characteristically whorled (3 at a node).......*Catalpa speciosa*

    4. Leaves characteristically opposite, seldom whorled.......................5

5. Leaves heart-shaped, large (6 to 12 in. long); exotic tree of city plantings, frequently escaped...................................*Paulownia tomentosa*

5. Leaves not heart-shaped, smaller..6

    6. Leaves entire or toothed, but not lobed....................................7

    6. Leaves both toothed and lobed...*Acer*

7. Leaves obscurely to strongly serrate or crenulate...............................10

7. Leaves strictly entire..8

    8. Leaves less than 2 in. long, ovate to elliptic; escaped shrub occasionally attaining tree form...................................*Ligustrum vulgare*

    8. Leaves more than 2 in. long; native...9

9. Leaf-blades typically oblong, 3 to 9 in. long; buds with several exposed scales; stipules absent....................................*Chionanthus virginicus*

9. Leaf-blades ovate to elliptic, seldom more than 4 in. long; vegetative buds elongate, with 2 valvate scales...*Cornus*

    10. Leaves distinctly serrate.....................................*Viburnum*

    10. Leaves obscurely serrate or crenulate toward the apex..................
....*Cornus florida*

Group E

Key to the genera with alternate compound leaves

1. Leaves pinnately compound (once-pinnate)...................................3

1. Leaves decompound (more than once-pinnate)...............................2

    2. Leaflets oval, about 5 pairs to each pinna; pod large, heavy (4-10 in. long); pith orange or salmon-colored...................*Gymnocladus dioica*

2. Leaflets one-sided, about 20-25 pairs to each pinna; pod flat, thin (2-4 in. long); pith white..*Albizzia julibrissin*

3. Leaves with glands on lower teeth, often with offensive odor when crushed ....*Ailanthus altissima*

3. Leaves without such glands.. 4

    4. Leaves with odor of green walnuts when crushed; pith chambered ....*Juglans*

    4. Leaves without walnut odor; pith not chambered......................... 5

5. Leaves with pulvini; fruit a legume.. 6

5. Leaves without pulvini; fruit not a legume..................................... 7

    6. Leaflets all opposite...*Robinia*

    6. Leaflets mostly alternate...*Cladrastis lutea*

7. Stipules or stipule scars present; buds red; vein-scars 3 or 5.................. ....*Sorbus americana*

7. Stipules absent; buds not red; vein-scars more numerous.................. 8

    8. Lateral buds partially or wholly concealed by petioles; fruit a small dry drupe; pith large..*Rhus*

    8. Lateral buds not concealed by petioles; fruit a nut, husk splitting along 4 lines; pith small, angled..*Carya*

Group F

Key to the genera with alternate simple leaves

1. Leaves strictly entire (never more than gently undulate).................... 2

1. Leaves variously toothed or lobed or both....................................10

    2. Leaves heart-shaped...*Cercis canadensis*

    2. Leaves not heart-shaped... 3

3. Leaves mostly more than 8 in. long.. 4

3. Leaves mostly less than 6 in. long... 5

    4. Twigs encircled by a stipular scar at each node.................*Magnolia*

    4. Without stipules ....*Asimina triloba*

5. Leaves characteristically clustered at tips of twigs with very short internodes; pith 5-angled; fruit an acorn..*Quercus*

5. Internodes not markedly shortened at branch tips; only one leaf at tip of each elongate vegetative branch; pith cylindrical or nearly so; fruit a drupe or berry.. 6

    6. Leaves with the upper 2 lateral veins strongly incurving.................. ....*Cornus alternifolia*

    6. Leaves strictly pinnately-veined to the tip..................................... 7

7. Twigs and leaves spicy-aromatic.......................................*Lindera benzoin*

7. Twigs and leaves not spicy-aromatic.. 8

    8. Leaves leathery, sweet to taste, often obscurely toothed................ ....*Symplocos tinctoria*

    8. Leaves neither leathery in texture nor sweet............................... 9

9. Pith diaphragmed but solid; vein-scars 3........................*Nyssa sylvatica*

9. Pith sometimes with cavities but not diaphragmed; vein-scar one........... .... *Diospyros virginiana*

    10. Leaf-blade usually averaging at least 1.5 times as long as broad....19

    10. Leaf-blade usually about as broad as long.................................11

11. Leaves more or less regularly toothed, but not lobed............................16
11. Leaves usually with a few conspicuous lobes, toothed or entire...........12
 12. Leaves bilaterally and symmetrically lobed.................................14
 12. Some leaves unlobed, others asymmetrically lobed........................13
13. Leaves coarsely serrate; fruit a multiple "berry"...........................*Morus*
13. Leaves not serrate; fruit a drupe..................................*Sassafras albidum*
 14. Leaf-tip truncate or broadly notched; leaves with one pair of broad,
  acute, lateral lobes..*Liriodendron tulipifera*
 14. Leaf-tip acuminate; leaves with main veins and lobes essentially
  palmate 15
15. Leaves star-shaped with deep notches between lobes, margin with fine,
 regular serrations...*Liquidambar styraciflua*
15. Leaves not star-shaped, with shallow sinuses; margins entire except
 for a few sinuate teeth...*Platanus occidentalis*
 16. Leaf-margins merely undulate or crenate; axillary buds stalked........
  ....*Hamamelis virginiana*
 16. Leaf-margins distinctly toothed; buds not stalked.......................17
17. Leaves all unlobed in our species, smooth above; sap not milky.......18
17. Trees usually with some irregularly lobed leaves but occasionally all
 unlobed; leaves usually somewhat harsh above; sap milky; fruit multiple
 ....*Morus*
 18. Leaves in 2 rows; pith cylindrical.........................*Tilia heterophylla*
 18. Leaves in more than 2 rows; pith 5-angled.........................*Populus*
19. Leaves characteristically clustered at tips of twigs, with very short internodes, prominently lobed or coarsely and regularly toothed; pith 5-
 angled; fruit an acorn...*Quercus*
19. Leaves not characteristically clustered at tips except on spur branches;
 if somewhat clustered, with glandular petioles; if somewhat lobed, less
 than 3 in. long; pith cylindrical; fruit not an acorn.............................20
 20. Sap milky; fruit multiple and fleshy; leaves ovate to cordate............
  ....occasional forms of *Morus*
 20. Sap not milky; fruit not multiple and fleshy; leaves ovate to
  lanceolate 21
21. Teeth of leaf-margins bristle-tipped...................................*Castanea*
21. Teeth of leaf-margins not bristle-tipped..............................22
 22. Leaves in 2 rows, more or less in one plane.......................23
 22. Leaves in more than 2 rows.......................................30
23. Leaves with 2 prominent lateral veins from base of blade; lateral buds
 appressed; pith typically chambered.......................................*Celtis*
23. Leaves otherwise; pith continuous.......................................24
 24. Leaves with main lateral veins dissipating into smaller veins before
  reaching the margin; fruit a small pome (apple-like); buds long
  and tapering ....*Amelanchier*
 24. Main lateral veins extending into teeth of leaf-margin; fruit not
  a pome 25
25. Terminal bud long and tapering, at least 4 times as long as broad;
 leaves coarsely serrate; fruit a bur with two triangular nuts..................
 ....*Fagus grandifolia*
25. Terminal buds less than 4 times as long as broad; leaves finely or
 doubly serrate 26

26. Most leaves bilaterally symmetrical or nearly so............................27
26. Most leaves decidedly lop-sided, especially at base, leaf-margins mostly doubly serrate ....*Ulmus*
27. Trunk and larger branches smooth, with fluted or projecting ridges, "muscular" in appearance; bud-scales in 4 rows.....*Carpinus caroliniana*
27. Trunk and larger branches without fluted or projecting ridges...........28
  28. Some lateral veins forked; bark longitudinally shredded; lenticels inconspicuous; fruit completely enclosed in a papery sac..................
  ....*Ostrya virginiana*
  28. Lateral veins unforked and continuous to leaf margin..................29
29. Bark relatively smooth except in very old trees; lenticels conspicuous, laterally elongated on larger branches and trunk; fruit winged, in cone-like clusters ....*Betula*
29. Bark ridged or scaly; lenticels inconspicuous; fruit a samara.......*Ulmus*
  30. Leaf-blades at least 4 times as long as broad..............31
  30. Leaf-blades less than 4 times as long as broad.............32
31. Bud with one exposed scale...*Salix*
31. Bud with about 6 exposed scales...................................*Prunus*
  32. Buds distinctly stalked; fruit a woody cone-like structure..............
  .... *Alnus serrulata*
  32. Buds not stalked; fruit otherwise................................33
33. Stipules or stipular scars present.....................................38
33. Neither stipules nor stipular scars present............................34
  34. Leaves leathery, sweet to taste; often obscurely toothed..................
  ....*Symplocos tinctoria*
  34. Leaves neither leathery in texture nor sweet....................35
35. Leaves distinctly sour to taste, margins ciliate; twigs greenish-red.........
  .... *Oxydendrum arboreum*
35. Leaves not sour; twigs brown or dark gray..........................36
  36. Twigs rusty-tomentose with persistent hairs; leaf-margins closely and sharply serrate....................................*Clethra acuminata*
  36. Twigs not tomentose; hairs, if present, scattered; leaf-margins not both closely and sharply serrate..............................37
37. Buds hairy, without scales; flowers axillary and solitary....*Stewartia ovata*
37. Buds glabrous, covered with scales; flowers in clusters..*Halesia carolina*
  38. Petioles with one or more glands near the blade; fruit a drupe........
  ....*Prunus*
  38. Petioles without glands..39
39. Wood of twigs yellowish and ill-scented; leaves obscurely toothed, with main lateral veins ending in margin.......................*Rhamnus caroliniana*
39. Wood of twigs neither yellowish nor ill-scented; leaves distinctly toothed with main lateral veins not extending to leaf-margin..............40
  40. Vein-scar one ....*Ilex*
  40. Vein-scars two or more..41
41. Younger twigs averaging less than ⅛ in. in diameter; leaves finely and regularly serrate ....*Amelanchier*
41. Younger twigs averaging more than ⅛ in. in diameter; leaves coarsely toothed or irregularly lobed...42
  42. Fruit apple-shaped ....*Malus*
  42. Fruit pear-shaped ....*Pyrus*

ALPHABETICAL LIST OF TREES
WITH KEYS TO SPECIES

Abies fraseri

Acer

1. Leaves compound..*A. negundo*
1. Leaves simple.. 2
  2. Buds with 4-8 scales apparent, essentially sessile; flowers in lateral clusters; trees of various habitats.................................... 4
  2. Buds with 2 valvate scales, distinctly stalked; flowers in terminal racemes; small trees of mountains.................................... 3
3. Twigs and buds glabrous; leaves finely serrate, with 3 main veins; bark striped with whitish lines................................*A. pensylvanicum*
3. Twigs and buds pubescent; leaves coarsely serrate, with 5 main veins; bark not striped..*A. spicatum*
  4. Leaves usually with 7 prominent veins from petiole; leaf-scars meeting; sap milky when evident; exotic trees frequent in city planting
....*A. platanoides*
  4. Leaves with 3 or 5 prominent veins from petiole; leaf-scars usually not meeting; sap not milky; native trees.................................... 5
5. Buds conical, exposed scales 6 or more.............................*A. saccharum*
5. Buds ovoid, usually 4 scales showing; flower buds rounded and collaterally multiple.. 6
  6. Lobes of leaves narrowed at the base; twigs ill-scented; bark flaking....
....*A. saccharinum*
  6. Lobes of leaves not narrowed at the base; twigs not ill-scented; bark tight, not flaking..*A. rubrum*

Aesculus octandra

Ailanthus altissima

Albizzia julibrissin

Alnus serrulata

Amelanchier

1. Leaves glabrous below; young leaves brownish-green.............*A. laevis*
1. Leaves pubescent below; young leaves whitish-green.............*A. arborea*

Aralia spinosa

Asimina triloba

Betula

1. Twigs with odor of wintergreen.. 2
1. Twigs without odor of wintergreen................................*B. nigra*
  2. Bark on young trees and branches peeling, yellowish; cone bracts ciliate; leaves cuneate or slightly heart-shaped at base................................
....*B. alleghaniensis*

2. Bark on young trees and branches cherry-like, tight; cone bracts glabrous; leaves heart-shaped or rounded at base.................*B. lenta*

Carpinus caroliniana

Carya

1. Buds with more than 6 overlapping scales; leaflets 3-9, the uppermost largest 3
1. Buds with 4-6 scales in pairs, meeting at edges; leaflets 7-17, usually lanceolate, often curved 2
   2. Leaflets 9-17, nut cylindric, longer than broad, shell thin, smooth; bark with flat, scaly, interlacing ridges..........................*C. illinoensis*
   2. Leaflets 7-13; nut somewhat flattened, about as broad as long, kernel bitter ....*C. cordiformis*
3. Larger terminal buds over ½ in. in length 4
3. Terminal buds smaller (less than ⅜ in.).................................... 6
   4. Twigs buff or orange-colored, glabrous; nut at least 1 in. long, shell thick; bark splitting off in long strips; leaflets 7-9; typically in bottom-lands or along streams...*C. laciniosa*
   4. Twigs brown or gray, often somewhat pubescent; nut and bark vari-ous; typically on uplands... 5
5. Twigs red-brown to gray with age; leaflets typically 5, terminal leaflet stalked; bark splitting off in long strips.............................*C. ovata*
5. Twigs bright brown to gray; leaflets typically 7-9, stellate-pubescent; ter-minal leaflet sessile or nearly so; bark tight......................*C. tomentosa*
   6. Leaflets typically 7..*C. pallida*
   6. Leaflets typically 5..*C. glabra*

*Castanea**

1. Leaves glabrous beneath; more than one nut in a bur; a large tree, most specimens of which are dead or dying..................................*C. dentata*
1. Leaves tomentose beneath; nut only one in a bur; a small tree seldom over 8 in. in diameter...*C. pumila*

Catalpa speciosa

Celtis

1. Leaf-blades seldom more than 2 in. in length; fruit dark orange-red, on stalks about as long as the petioles; small tree..*C. tenuifolia* var. *georgiana*
1. Leaf-blades usually more than 2 in. in length; fruit on stalks longer than the petioles; becoming large trees.................................. 2
   2. Leaf-blades entire, or toothed toward the apex; bark light gray, with corky warts...*C. laevigata*
   2. Leaf-blades strongly toothed to well below the middle..*C. laevigata* var. *smallii*

Cercis canadensis

   * *C. mollissima*, a rare exotic species, grows in Cades Cove.

Chionanthus virginicus

Cladrastis lutea

Clethra acuminata

Cornus

1. Leaves always opposite; fruit red; inflorescence with large white bracts....
....*C. florida*
1. Leaves irregularly alternate; fruit blue; inflorescence without white bracts
....*C. alternifolia*

Crataegus

Ten species of this difficult and confused genus have been credited to Great Smoky Mountains National Park. Flower and fruit characters are required for identification. The authors have attempted no key.

Diospyros virginiana

Fagus grandifolia

Fraxinus

1. Twigs and leaves glabrous...*F. americana*
1. Twigs pubescent; leaves more or less pubescent.............*F. pennsylvanica*

Gleditsia triacanthos

Gymnocladus dioica

Halesia carolina

Hamamelis virginiana

Ilex

1. Leaves evergreen..*I. opaca*
1. Leaves deciduous.. 2
   2. Leaves averaging 3 in. long; flowers and fruits short-stalked..............
   ....*I. montana*
   2. Leaves averaging less than 3 in. long; flowers and fruits long-stalked
   ....*I. longipes*

Juglans*

1. Pith chocolate-colored; leaf-scars with a downy cross-line at top, not notched; fruit longer than broad, hull sticky-glandular............*J. cinerea*
1. Pith tan; leaf-scars without a downy ridge at top, notched; fruit essentially spheroidal, hull not glandular..*J. nigra*

   * *J. mandshurica,* a rare exotic species, reported from Cades Cove.

Juniperus virginiana

Kalmia latifolia

Ligustrum vulgare

Lindera benzoin

Liquidambar styraciflua

Liriodendron tulipifera

Maclura pomifera

Magnolia

1. Leaves deciduous, not leathery.. 2
1. Leaves evergreen, leathery..*M. grandiflora*
   2. Leaves cordate at the base.. 4
   2. Leaves not cordate at the base.. 3
3. Leaves 6-10 in. long..*M. acuminata*
3. Leaves 18-20 in. long...*M. tripetala*
   4. Leaves strongly auriculate, not whitened beneath, 10-12 in. long;
      petals 10-16 in. long...*M. fraseri*
   4. Leaves not strongly auriculate, pale to nearly white beneath, 20-30 in.
      long; petals 6 in. long..*M. macrophylla*

Malus

1. Branches usually armed with hard, sharp lateral spurs......*M. angustifolia*
1. Branches unarmed..*M. pumila*

Morus

1. Leaves harsh above, more or less tomentose below, infrequently lobed;
   fruit red...*M. rubra*
1. Leaves smooth on both sides, usually lobed; fruit white...............*M. alba*

Nyssa sylvatica

Ostrya virginiana

Oxydendrum arboreum

Paulownia tomentosa

Picea

1. Leaves averaging 2/5 in. long..*P. rubens*
1. Leaves averaging 4/5 in. long..*P. abies*

Pieris floribunda

Pinus

1. Leaves characteristically 5 in a bundle......................................*P. strobus*

1. Leaves 2 or 3 in a bundle 2
    2. Leaves 8-18 in. long .... *P. palustris*
    2. Leaves much less than 8 in. long 3
3. Leaves characteristically 2 in a bundle; or in both twos and threes 4
3. Leaves characteristically 3 in a bundle .... *P. rigida*
    4. Cones commonly asymmetrical, often more than 2½ in. in length,
       with very stout prickles .... *P. pungens*
    4. Cones usually symmetrical, with slender prickles, less than 2½ in.
       in length 5
5. Branches nearly smooth; leaves twisted, usually less than 2 in. long. in
   twos .... *P. virginiana*
5. Branches scaly; leaves not twisted, usually 2½ to 5 in. long, usually in
   both twos and threes .... *P. echinata*

Platanus occidentalis

Populus
1. Fastigiate (with upright branches) .... *P. nigra* var. *italica*
1. Not fastigiate 2
    2. Teeth small, more than 14 on each side 4
    2. Teeth large, less than 14 on each side of the leaf blade 3
3. Petioles averaging over 2 in. in length; twigs and leaves essentially
   glabrous .... *P. grandidentata*
3. Petioles averaging less than 2 in.; twigs and leaves white tomentose ......
   .... *P. alba*
    4. Petioles smooth .... *P. deltoides*
    4. Petioles hairy .... X *P. gileadensis*

Prunus
    (A difficult group when without fruit and flower characters)
1. Terminal-bud typically present 5
1. Terminal-bud typically absent, represented by a scar 2
    2. Buds elongate, longer than thick 4
    2. Buds scarcely longer than thick 3
3. Leaves usually 2-4 in. long; calyx lobes glandular .... *P. munsoniana*
3. Leaves mostly 1-2 in. long; calyx lobes without glands ..... *P. angustifolia*
    4. Leaves thin, lustrous, acute or acuminate, crenate-dentate ..*P. hortulana*
    4. Leaves dull, dark green, abruptly pointed at apex, sharply serrate ......
       .... *P. americana*
5. Twigs green or red .... *P. persica*
5. Twigs reddish-brown or gray 6
    6. Buds averaging 3/16 in. long; flowers in elongate clusters 7
    6. Buds ⅛ in. long or less; flowers not in elongate clusters 8
7. Leaves slenderly pointed; petals longer than wide; sepals adhering to the
   fruits .... *P. serotina*
7. Leaves abruptly pointed; petals often wider than long; sepals early
   deciduous .... *P. virginiana*
    8. Buds ⅛ in. long or less .... *P. pensylvanica*
    8. Buds 3/16 to ¼ in. long (escaped, edible cherries) 9

9. Buds glossy, ovoid-fusiform...*P. avium*
9. Buds duller or darker, round-ovoid...*P. cerasus*

Pyrus communis

Quercus
1. Leaves characteristically entire (unlobed and untoothed)....*Q. imbricaria*
1. Leaves characteristically lobed, toothed, or both............................ 2
   2. Leaves broadest near the tip (about 1/6-¼ from the apex); not conspicuously lobed or toothed...................................*Q. marilandica*
   2. Leaves broadest nearer the middle, with conspicuous teeth or lobes 3
3. Leaves distinctly lobed.. 5
3. Leaves with coarse teeth or scalloped but not distinctly lobed............ 4
   4. Leaves coarsely sinuate-toothed or with irregular very shallow lobes, usually with 6-8 pairs of lateral veins, not all ending in an acute point; acorns on stalks approximately 1 in. long............................*Q. bicolor*
   4. Leaves mostly with more than 9 pairs of lateral veins most of which end in a point; acorns on short stalks...................................*Q. prinus*
5. Lobes of leaves without bristle-tips... 6
5. Lobes of leaves with bristle-tips.. 7
   6. Leaves glaucous and glabrous beneath at maturity.................*Q. alba*
   6. Leaves densely gray-pubescent beneath............................*Q. stellata*
7. Mature leaves more or less pubescent on the whole undersurface....... 8
7. Mature leaves smooth beneath except for tufts of hairs in the major vein-angles 9
   8. Leaves brownish or rusty pubescent beneath, lobes not curved, frequently wider toward the end...................................*Q. velutina*
   8. Leaves grayish or yellowish pubescent beneath, lobes generally curved and widest at the base...*Q. falcata*
9. Lateral lobes of leaves not decidedly longer than the width of the undivided portion of the blade; leaves dull, 7-11 lobed; acorn cup saucerlike ....*Q. rubra*
9. Lateral lobes of leaves decidedly longer than the width of the undivided portion of the blade; leaves lustrous, 5-9 lobed....................*Q. coccinea*

Rhamnus caroliniana

Rhododendron
1. Leaf-blades about 3.5 times as long as broad, tapering at base, green to brownish beneath...*R. maximum*
1. Leaf-blades about 2 times as long as broad, rounded at base, whitish beneath..*R. catawbiense*

Rhus
1. Leaf scars U-shaped; rachis winged.................................*R. copallina*
1. Leaf scars C-shaped or broadly crescent-shaped; rachis not winged...... 2
   2. Stems glabrous or nearly so; twigs often 3-sided...................*R. glabra*
   2. Stems hairy, concealing the lenticels; twigs rounded............*R. typhina*

Robinia pseudoacacia

Salix
1. Branchlets strongly drooping.................................*S. babylonica*
1. Branchlets not strongly drooping... 2

2. Leaves whitish beneath..*S. alba*
2. Leaves green beneath..*S. nigra*

Sassafras albidum

Sorbus americana

Stewartia ovata

Symplocos tinctoria

Thuja
1. Branchlets horizontally flattened; cone-scales not thick......*T. occidentalis*
1. Branchlets vertically flattened; cone-scales thick....................*T. orientalis*

Tilia heterophylla

Tsuga canadensis

Ulmus
1. Leaves mostly less than 2¾ in. long; twigs usually with corky wings....
....*U. alata*
1. Leaves usually more than 2¾ in. long; twigs without corky wings...... 2
   2. Bud-scales coated with rusty hairs; leaves very rough above; pedicels
     short; fruit not ciliate; inner bark mucilaginous......................*U. rubra*
   2. Bud-scales without rusty hairs; leaves relatively smooth above; pedicels
     slender, drooping; fruit ciliate; inner bark not mucilaginous..............
     .... *U. americana*

Vaccinium arboreum

Viburnum
1. Buds short and broad, dark red-scurfy..................................*V. rufidulum*
1. Buds long and slender, gray or reddish-brown..................*V. prunifolium*

Zanthoxylum americanum

KEY TO SHRUBS

1. Leaves evergreen 2
1. Leaves deciduous 14
   2. Leaves opposite 3
   2. Leaves alternate 5
3. Plants parasitic, growing on limbs of deciduous trees............................
....*Phoradendron flavescens*
3. Plants rooted in soil.. 4
   4. Leaves leathery, rather closely clustered............*Buxus sempervirens*
   4. Leaves membranous, loosely arranged in 2 rows....*Ligustrum vulgare*
5. Leaves grass-like 6
5. Leaves not grass-like 8
   6. Plants short; leaves tightly clustered......................*Yucca smalliana*

158 KEY TO SHRUBS

    6. Plants bamboo-like 7
  7. Leaves not over 1 in. wide; several branches appearing at a node........
    ....*Arundinaria tecta*
  7. Leaves usually 1½ in. wide; one branch at a node....*Pseudosasa japonica*
    8. Plants usually less than 6 in. high.................................... 9
    8. Plants usually over 1 ft. high.....................................10
  9. Leaves lance-shaped, toothed, with white central line........................
    ....*Chimaphila maculata*
  9. Leaves rounded, with smooth margin, uniformly green........................
    ....*Gaultheria procumbens*
    10. Leaves small, ½ in. or less in length........*Leiophyllum buxifolium*
        var. *hugeri*
    10. Leaves usually 2 in. or more in length............................11
 11. Plants arching and sprawling....................*Leucothoë fontanesiana*
 11. Plants erect..12
    12. Leaves usually over 4 in. in length or punctate with brown scales
    on the undersurface...*Rhododendron*
    12. Leaves usually less than 3 in. in length, never punctate............13
 13. Young twigs hairy; flowers always white, small, vase-shaped in elon-
    gated terminal clusters....................................*Pieris floribunda*
 13. Young twigs usually smooth; flowers over ½ in. in diameter, broadly
    bell-shaped, in flattened terminal clusters.................*Kalmia latifolia*
    14. Leaves compound..15
    14. Leaves simple 25
 15. Leaf-blades of 3 leaflets..16
 15. Leaf-blades of 5 or more leaflets..18
    16. Leaves opposite..................................*Staphylea trifolia*
    16. Leaves alternate 17
 17. Stems without thorns or spines.........................*Cytisus scoparius*
 17. Stems with thorns or spines.......................................*Rubus*
    18. Stems without thorns or spines.....................................19
    18. Stems with thorns or spines..22
 19. Stems with bright yellow interior.................*Xanthorhiza simplicissima*
 19. Stems without bright yellow interior.......................................20
    20. Leaves opposite ....*Sambucus*
    20. Leaves alternate 21
 21. Leaflets serrate or toothed.......................................*Rhus*
 21. Leaflets with entire margins...................................*Amorpha*
    22. Leaves decompound, composed of innumerable leaflets, the largest
    leaf in the park...*Aralia spinosa*
    22. Leaves once-pinnately compound............................... 23
 23. Leaves when crushed strongly aromatic with lemon odor
    ....*Zanthoxylum americanum*
 23. Leaves slightly, if at all, aromatic..24
    24. Leaflets with serrate margins.....................................*Rosa*
    24. Leaflets with entire margins...................................*Robinia*
 25. Leaves opposite 26
 25. Leaves alternate..43
    26. Leaves lobed.......................................*Acer spicatum*
    26. Leaves sometimes toothed but never lobed........................27
 27. Leaf-blades cordate, heart-shaped (*Viburnum alnifolium* may be sought
    here)...*Syringa vulgaris*

27. Leaf-blades not heart-shaped..28
    28. Leaves when crushed with spicy fragrance..............................
    ....*Calycanthus floridus* var. *laevigatus*
    28. Leaves when crushed without spicy fragrance..............................29
29. Young twigs rather persistently green................................30
29. Young twigs soon losing their green color....................31
    30. Main leaf-veins not heavy, straight to slightly curved.....*Euonymus*
    30. Main leaf-veins prominent, curving toward the tip of leaf..............
    .... *Cornus*
31. Young twigs definitely reddish..32
31. Young twigs gray or brown..33
    32. Stem enlarged at nodes; fruits fleshy in loose clusters......................
    .... *Cornus amomum*
    32. Stem not enlarged at nodes; fruits dry in a tight spherical cluster....
    .... *Cephalanthus occidentalis*
33. Leaves usually 1½ in. or less in length.........................34
33. Leaves often 2 in. or more in length............................36
    34. Leaves less than twice as long as wide....*Symphoricarpos orbiculatus*
    34. Leaves often more than twice as long as wide......................35
35. Leaves narrow, not over 1 in. long.....................*Ascyrum hypericoides*
35. Leaves broad, usually over 1½ in. long..................*Lonicera canadensis*
    36. Leaves entire...37
    36. Leaves serrate to dentate...38
37. Leaves lance-shaped, with distinctly thickened margins...........*Viburnum*
37. Leaves broadly oval, without thickened margins..*Chionanthus virginicus*
    38. Flowers numerous, small, densely clustered in the leaf-angles, be-
    coming magenta-colored fruits.........................*Callicarpa americana*
    38. Flowers loosely arranged in the leaf-angles....................39
39. Bark breaking loose in shreds the second year....................40
39. Bark tight for several years...41
    40. Leaf-margins evenly serrate.....................................*Hydrangea*
    40. Leaf-margins irregularly serrate to dentate...................*Philadelphus*
41. Bark on young twigs with numerous, conspicuous, corky, pimple-like
    lenticels; flowers in leaf-angles.........................*Forsythia*
41. Bark on young twigs with less conspicuous lenticels; flowers in terminal
    clusters except in *Diervilla*..42
    42. Flower-clusters terminal ....*Viburnum*
    42. Flowers not all terminal................................*Diervilla sessilifolia*
43. Leaves spicy, aromatic when crushed.............................*Lindera benzoin*
43. Leaves without any spicy odor..44
    44. Leaf-margins entire, occasionally minutely serrulate...................45
    44. Leaf-margins serrate or dentate.......................................54
45. Leaves broad, heart-shaped or reniform.....................*Cercis canadensis*
45. Leaves lance-shaped 46
    46. Leaves large, 6-10 in. long, broadest in the upper half..............
    .... *Asimina triloba*
    46. Leaves smaller, broadest near or below the middle................47
47. Middle vein of leaf conspicuously thickened at the tip.................48
47. Middle vein of leaf not much thickened at the tip.................49
    48. Mature leaves averaging 1 in. or less in length........*Menziesia pilosa*
    48. Mature leaves averaging more than 1 in. long...........*Rhododendron*

49. Leaves with minute, light-reflecting spots or glands on the undersurface
....*Gaylussacia*
49. Leaves without such spots..50
    50. Leaves broadest below the middle, usually near the base, occasionally slightly serrate................................*Celtis tenuifolia*
    50. Leaves broadest near the middle................................51
51. Leaves averaging more than 1 in. wide................................52
51. Leaves averaging less than 1 in. wide................................53
    52. Leaf-surface marked by a conspicuous network of protruding veins
....*Pyrularia pubera*
    52. Leaf-surface relatively smooth; veins not conspicuously protruding
....*Symplocos tinctoria*
53. Leaves averaging less than 4 times as long as wide; fruit a fleshy berry
....*Vaccinium*
53. Leaves averaging usually more than 4 times as long as wide; fruit a dry
capsule filled with cottony seeds................................*Salix*
    54. Bark conspicuously exfoliating................................55
    54. Bark slightly or not at all exfoliating................................56
55. Pieces of loose bark broad, red................................*Clethra acuminata*
55. Pieces of loose bark narrow, brown................*Physocarpus opulifolius*
    56. Twigs with spines................................57
    56. Twigs without spines................................59
57. Leaves lobed (*Hibiscus syriacus* may be sought here)................*Ribes*
57. Leaves unlobed, lance-shaped................................58
    58. Twigs with clustered spines................................*Berberis vulgaris*
    58. Twigs with spines appearing singly................*Chaenomeles lagenaria*
59. Leaves large, over 4 in. wide, broadly lobed, shaped like a maple leaf
....*Rubus odoratus*
59. Leaves not as above................................60
    60. Leaves averaging about 4 in. long, evenly toothed with a weak
spine at the tip of each tooth................................*Castanea pumila*
    60. Leaves averaging less than 4 in. or if that long not evenly
toothed 61
61. Leaves rounded, margin wavy to coarsely toothed; leaf-tip obtuse (See
also *Ribes*) 62
61. Leaf-tip acute to acuminate................................63
    62. Leaves seldom less than 2 in. long; flowers yellow, in autumn........
....*Hamamelis virginiana*
    62. Leaves averaging less than 2 in. long; flowers white, in spring........
....*Fothergilla major*
63. Shrubs with cone-like clusters of female flowers or fruits....................
....*Alnus serrulata*
63. Shrubs without such clusters................................64
    64. Leaves with long slender tips about ⅓ the length of the leaf.........
....*Kerria japonica*
    64. Leaves with acute but not long and slender tips................65
65. Male flowers borne in catkins................................*Corylus*
65. Flowers not borne in catkins................................66
    66. Flowers large (hollyhock-like), 2 in. or more wide............*Hibiscus*
    66. Flowers smaller, 1 in. or less wide................................67
67. Plants seldom over 3 ft. high; flowers small, white................68
67. Plants often over 3 ft. high................................69

68. Shrub arising from a red root; flowers not urn-shaped..................
.... *Ceanothus americanus*
68. Roots not red; flowers urn-shaped.........................*Lyonia ligustrina*
69. Flowers borne along the stem next to the attachment of the leaves....*Ilex*
69. Flowers borne at the ends of twigs or branches...................................70
    70. Flowers borne in slender, rather flexuous clusters........................71
    70. Flowers borne in short, rather flat-topped clusters.......................73
71. Leaf-base acute; buds inconspicuous, blunt; flower-cluster spike-like....72
71. Leaf-base rounded; buds long, sharply-pointed; flower-cluster not spike-
like.. *Amelanchier sanguinea*
    72. Petals separate; fruit an elongated capsule; growing in swampy soil
....*Itea virginica*
    72. Petals fused; fruit a squat, globular capsule; growing on ridge tops
    or slopes...*Leucothoë recurva*
73. Leaves with regular, small, single teeth..*Pyrus*
73. Leaves with irregular and double teeth (small teeth on larger teeth).......
.... *Spiraea* spp.

ALPHABETICAL LIST OF SHRUBS
WITH KEYS TO SPECIES

Acer spicatum

Alnus serrulata

Amelanchier sanguinea

Amorpha
1. Leaves usually pubescent; calyx usually over 1/12 in. in length..............
.... *A. fruticosa*
1. Leaves almost glabrous; calyx 1/12 in. or less in length............*A. glabra*

Aralia spinosa

Arundinaria tecta

Ascyrum hypericoides

Asimina triloba

Berberis vulgaris

Buxus sempervirens

Callicarpa americana

Calycanthus floridus var. *laevigatus*

Castanea pumila

Ceanothus americanus

Celtis tenuifolia

Cephalanthus occidentalis

Cercis canadensis

Chaenomeles lagenaria

Chimaphila maculata

Chionanthus virginicus

Clethra acuminata

Cornus amomum

Corylus
1. Leaf-stalks with glandular hairs; fruits with a short covering..................
.... *C. americana*
1. Leaf-stalks without glandular hairs; fruits with a longer beaked covering
....*C. cornuta*

Cytisus scoparius

Diervilla sessilifolia

Euonymus
1. With the aspect of a small tree, having a larger central trunk with rough
   bark; fruit smooth...*E. atropurpureus*
1. With arching or trailing, slender, green stems; fruits rough................. 2
   2. Plants trailing, usually less than 1½ ft. high; upper leaves obovate....
   ....*E. obovatus*
   2. Plants arching (occasionally erect), up to 6 ft. tall; leaves widest in the
      middle ....*E. americanus*

Forsythia spp.

Fothergilla major

Gaultheria procumbens

Gaylussacia
1. Plants of dry areas; leaves small, leathery, obtuse, with copious reflecting
   dots... *G. baccata*
1. Plants of shady, moist areas; leaves acute, rather thin, with scattered
   reflecting dots...*G. ursina*

Hamamelis virginiana

Hibiscus
1. Plants almost herbaceous, in swampy soil......*H. palustris* var. *oculiroseus*
1. Plants definitely shrubby, persisting around old plantings.......*H. syriacus*

Hydrangea
1. Leaves green beneath...*H. arborescens*
1. Leaves silvery-white beneath......................................*H. radiata*

Ilex
1. Leaves averaging 3 in. long; flowers and fruits short-stalked....*I. montana*
1. Leaves averaging less than 3 in. long; flowers and fruits long-stalked....
....*Ilex longipes*

Itea virginica

Kalmia latifolia

Kerria japonica

Leiophyllum buxifolium var. *hugeri*

Leucothoë fontanesiana

Leucothoë recurva

Ligustrum vulgare

Lindera benzoin

Lonicera canadensis

Lyonia ligustrina

Menziesia pilosa

Philadelphus
1. Flowers in clusters of 3 or less; styles separate above in flowers and
   fruits ....*P. pubescens*
1. Flowers in clusters of 5 or more; styles united throughout in flowers and
   fruits 2
      2. Hypanthium pubescent; leaves scabrous-hirsute above.......*P. hirsutus*
      2. Hypanthium glabrous; leaves strigose-pilose above.........*P. sharpianus*

Phoradendron flavescens

Physocarpus opulifolius

Pieris floribunda

Pseudosasa japonica

Pyrularia pubera

Pyrus
1. Younger branches and lower sides of the leaves slightly or not at all
   hairy...*P. melanocarpa*
1. Younger branches and usually the lower sides of leaves hairy............... 2
      2. Fruits red, about 1/5 in. in diameter...............................*P. arbutifolia*
      2. Fruits purplish-black, about 2/5 in. in diameter.............*P. prunifolia*

Rhododendron
1. Leaves evergreen... 2
1. Leaves deciduous... 4

2. Plants usually on exposed sites; leaves averaging less than 3 in. long, with brown scales on the undersurface...*R. minus*
2. Plants on various sites; leaves averaging over 4 in. long, without scales 3
3. Leaves acute at the base, seldom whitish beneath.................*R. maximum*
3. Leaves rounded at the base, often whitish beneath.............*R. catawbiense*
    4. Flowers appearing with or before the leaves.................................... 5
    4. Flowers appearing after the leaves... 6
5. Lobes of the corolla about as long as the fused basal portion.................. ....*R. nudiflorum*
5. Lobes of the corolla clearly shorter than the fused basal portion............ ....*R. canescens*
    6. Flowers pale orange to red.....................................*R. calendulaceum*
    6. Flowers white or pink.. 7
7. Stamens about twice as long as the corolla-tube...................*R. viscosum*
7. Stamens usually much more than twice as long as the corolla-tube.......... .... *R. arborescens*

Rhus (see Key to Trees)

Ribes
1. Stems with scattered spines.. 2
1. Stems without spines... 3
    2. Twigs and fruit bristly, in addition to the spines at the base of the leaves..*R. cynosbati*
    2. Twigs with scattered spines at bases of some leaves; fruit smooth ....*R. rotundifolium*
3. Stems sprawling, with an unpleasant odor where bruised; flowers not yellow, ill-scented...*R. glandulosum*
3. Stems more or less erect, without unpleasant odor; flowers yellow, pleas-antly scented ....*R. odoratum*

Robinia* (also see text pages 89-91)

Rosa

Rubus
1. Leaves lobed but simple..*R. odoratus*
1. Leaves compound 2
    2. Fruits thimble-shaped, at maturity separating from a central dome.... 3
    2. Fruits not thimble-shaped, tightly adhering to a central axis............. 4
3. Stems densely hairy..*R. idaeus* var. *canadensis*
3. Stems smooth except for scattered spines..........................*R. occidentalis*
    4. Stems prostrate or low-arching, seldom reaching 1 ft. above the soil ....Dewberries*
    4. Stems arching to erect, usually reaching 2 ft. above the soil............. ....Blackberries*

Salix
1. Leaf-margins entire or nearly so...*S. humilis*
1. Leaf-margins finely serrate.. 2

* These are so difficult to identify that the specialists disagree.

2. Aments appearing with the leaves, 1 1/6 to 4 in. long; flower scales yellowish, deciduous; stamens 5-8..................................*S. caroliniana*
2. Aments appearing before the leaves, ⅝-1 in. long; flower scales dark brown, persistent; stamens 2...*S. sericea*

Sambucus
1. Pith white; flowers white in a flat cluster; mature fruits black.................
....*S. canadensis*
1. Pith brown; flowers cream in a pyramidal cluster; mature fruits red........
....*S. pubens*

Spiraea spp.

Staphylea trifolia

Symphoricarpos orbiculatus

Symplocos tinctoria

Syringa vulgaris

Vaccinium
1. Leaves leathery, rather oval, glossy; fruits black and dry, inedible...........
....*V. arboreum*
1. Leaves membranous; fruits fleshy... 2
  2. Leaves and young twigs hairy.. 3
  2. Leaves and young twigs smooth.. 4
3. Cylindrical flowers and fruits densely hairy, without a tiny leaf at the base of the flower-stalk...*V. hirsutum*
3. Flaring, bell-shaped flowers, and fruits relatively smooth, with a tiny leaf at the base of each flower-stalk.....................................*V. stamineum*
  4. Leaves slenderly acute to acuminate; corolla with 4 slender recurving petals; flowers and fruits occurring singly.................*V. erythrocarpum*
  4. Leaves merely acute; petals fused into an urn-shaped corolla; flowers and fruits occurring in clusters..Blueberries*

Viburnum
1. Leaves about as broad as long... 2
1. Leaves longer than wide.. 4
  2. Leaf-blade unlobed, cordate with a single prominent vein....................
....*V. alnifolium*
  2. Leaf-blade lobed like a maple leaf, with 3 prominent veins................... 3
3. A cultivated species; marginal flowers much larger than the rest of the flower-cluster ....*V. opulus*
3. A native species; all flowers of the cluster essentially similar....................
....*V. acerifolium*
  4. Clusters of flowers or fruits with long stalks.................*V. cassinoides*
  4. Clusters of flowers or fruits essentially without stalks....................... 5
5. Petioles broadly flattened, with dense, scurfy, red hairs........*V. rufidulum*
5. Petioles narrow, with few or no hairs...............................*V. prunifolium*

Xanthorhiza simplicissima

* These are so difficult to identify that the specialists disagree.

Yucca smalliana

Zanthoxylum americanum

KEY TO WOODY VINES*

1. Leaves simple 11
1. Leaves compound 2
    2. Leaves palmately compound...................*Parthenocissus quinquefolia†*
    2. Leaves pinnately or trifoliolately compound................................. 3
3. Leaves evergreen... 4
3. Leaves deciduous... 7
    4. Climbing vine with leaflets in pairs.....................*Bignonia capreolata*
    4. Vines trailing, or at least rarely over 6 in. tall...................... 5
5. Leaves alternate, hairy..*Epigaea repens*
5. Leaves opposite, smooth... 6
    6. Plants prostrate; leaves ovate..................................*Mitchella repens*
    6. Plants arching; leaves ovate-lanceolate...........................*Vinca minor*
7. Leaves usually with 7 or more leaflets.. 8
7. Leaves usually with 5 or fewer leaflets....................................... 9
    8. Leaflets rather strongly serrate.............................*Campsis radicans*
    8. Leaflets never serrate......................................*Wisteria frutescens*
9. Stems without roots between leaves...10
9. Stems with roots between the leaves; no part of the leaf twining...........
    ....*Rhus radicans*
    10. With appendages at the base of the leaf-stalk; no part of leaf
        twining..*Pueraria lobata*
    10. Without appendages at the base of the leaf-stalk; parts of leaves
        often twining ....*Clematis*
11. Leaves essentially evergreen...12
11. Stems without thorns or spines; leaves deciduous...........................13
    12. Stems with thorns or spines....................................*Smilax*
    12. Stems without thorns or spines...............................*Lonicera*
13. Leaf-margins without teeth although rarely broadly lobed..................14
13. Leaf-margins toothed 15
    14. High-climbing, over 15 ft. in length.................*Aristolochia durior*
    14. Sprawling, usually less than 10 ft. in length.........*Cocculus carolinus*
15. Leaves clearly longer than broad, finely toothed.......................*Celastrus*
15. Leaves about as broad as long, coarsely toothed............................16
    16. Plants rarely over 10 ft. long; leaves coarsely and bluntly toothed
        (almost lobed)...*Menispermum canadense*
    16. Plants high climbing; leaves always sharply toothed....................17
17. Bark tight..18
17. Bark loose and shredding..*Vitis*
    18. Pith brown in older stems.......................*Vitis rotundifolia*
    18. Pith always white............................*Ampelopsis cordata*

   * *Gaultheria procumbens* and *Euonymus obovatus* may occasionally be
sought here (see key to shrubs).

   † If hairy, the var. *hirsuta.*

ALPHABETICAL LIST OF WOODY VINES
WITH KEYS TO SPECIES

Ampelopsis cordata

Aristolochia durior

Bignonia capreolata

Campsis radicans

Celastrus
1. Leaves usually twice as long as wide..*C. scandens*
1. Leaves about 1½ times as long as wide................................*C. orbiculatus*

Clematis
1. Leaves thin and membranous, with 3 leaflets......................*C. virginiana*
1. Leaves thicker and often rather leathery, with 2-5 leaflets.................. 2
    2. Leaves usually with 5 leaflets; flowers flat, white............*C. paniculata*
    2. Leaves with 2-4 leaflets; flowers urn-shaped, purplish...........*C. viorna*

Cocculus carolinus

Epigaea repens

Lonicera
1. None of the opposite leaves fused; flowers white...................*L. japonica*
1. Opposing leaves beneath the flowers fused to each other at the base;
    flowers red or yellow...*L. sempervirens*

Menispermum canadense

Mitchella repens

Parthenocissus quinquefolia

Pueraria lobata

Rhus radicans

Smilax
1. Leaves leathery, shiny, and tough throughout; stems with stout thorns....
    .... *S. rotundifolia*
1. Leaves thin; stems often densely spiny but seldom with stout thorns.... 2
    2. Leaves white beneath..*S. glauca*
    2. Leaves green beneath.....................................*S. tamnoides* var. *hispida*

Vinca minor

Vitis
1. Leaf-blades small, glossy on both surfaces, rounded, sometimes wider
    than long; bark tight, not exfoliating; pith continuous throughout stem....
    ....*V. rotundifolia*

1. Leaf-blades longer than wide, not glossy on both surfaces; pith discontinuous; bark exfoliating in loose shreds... 2
   2. Leaves with permanently hairy undersurfaces....................................3*
   2. Leaves with undersurfaces smooth with age or with hairs only on the veins or in their angles.. 4

3. Hairs rusty or reddish..*V. aestivalis*
3. Hairs grayish...*V. cinerea*
   4. Leaves when older essentially without hairs except in the vein-angles on the undersurface..*V. vulpina*
   4. Leaves hairy beneath *on the veins* and in the vein-angles................ 5
5. Tendrils often produced on 3 or more consecutive nodes........*V. labrusca*
5. Tendrils never produced on 3 consecutive nodes.................*V. baileyana*

Wisteria frutescens

   * See also *V. labrusca.*

References Cited ——————————

Ammons, Nelle
    1950.   Shrubs of West Virginia. *W. Va. Univ. Bull.* Series 50, No. 12-4. 127 pp.

Ayres, H. B. and W. W. Ashe
    1905.   The Southern Appalachian Forests. *U. S. Geol. Surv. Professional Paper* No. 37, Washington, D. C. 291 pp.

Bailey, L. H.
    1934a.   The Species of Grapes Peculiar to North America. *Gentes Herb.* 3:149-244.
    1934b.   Certain Northern Blackberries. *Gentes Herb.* 3:247-271.
    1937.   *The Standard Cyclopedia of Horticulture.* 3 vols. Macmillan Co., New York. 3639 pp.
    1944.   The Genus *Rubus* in North America. *Gentes Herb.* 5:507-588.
    1945.   The Genus *Rubus* in North America. *Gentes Herb.* 5:591-856.

Baldwin, S. G.
    1948a.   Photographing Big Trees in the Smokies. *Jour. Biol. Photographic Assoc.* 17: No. 1 (Sept.).
    1948b.   Big Trees of the Great Smokies. *Southern Lumberman,* Dec. 15, pp. 172-178.

Bartram, W.
    1791.   *Travels through North and South Carolina, Georgia, East and West Florida . . .* (1928 printing). Macy-Masius, New York. 522 pp.

Billings, W. D., S. A. Cain, and W. B. Drew
    1937.   Winter Key to the Trees of Eastern Tennessee. *Castanea (Jour. So. Appal. Bot. Club)* 2:29-44.

Billings, W. D. and A. F. Mark
    1957.   Factors Involved in the Persistence of Montane Treeless Balds. *Ecology* 38:140-142.

Braun, E. L.
    1950.   *Deciduous Forests of Eastern North America.* Blakiston Co., Philadelphia. 596 pp.
    1961.   *The Woody Plants of Ohio.* Ohio State Univ. Press. Columbus. 362 pp.

Brown, C. A.
1945.   Louisiana Trees and Shrubs. *Bull. No. 1,* La. Forestry Commission, Baton Rouge. 262 pp.

Buckley, S. B.
1859.   Mountains of North Carolina and Tennessee. *Amer. Jour. Sci. and Arts* (2nd series) 27:286-294.

Cain, S. A.
1930a.   Certain Floristic Affinities of the Trees and Shrubs of the Great Smoky Mountains and Vicinity. *Butler Univ. Bot. Studies* 1:129-150.

1930b.   An Ecological Study of the Heath Balds of the Great Smoky Mountains. *Butler Univ. Bot. Studies* 1:177-208.

1931.   Ecological Studies of the Vegetation of the Great Smoky Mountains of North Carolina and Tennessee. I. Soil reaction and plant distribution. *Bot. Gaz.* 91:22-41.

1935.   Ecological Studies of the Vegetation of the Great Smoky Mountains. II. The quadrat method applied to sampling spruce and fir forest types. *Amer. Midland Nat.* 16:566-584.

1936.   Ecological Work on the Great Smoky Mountains Region. *Castanea (Jour. So. Appal. Bot. Club)* 1:25-32.

1940.   An Interesting Behavior of Yellow Birch in the Great Smoky Mountains. *Chicago Nat.* 3:20-21.

1943.   The Tertiary Character of the Cove Hardwood Forests of the Great Smoky Mountains National Park. *Bull. Torrey Bot. Club* 70:213-235.

1944.   *Foundations of Plant Geography.* Harper, New York. 556 pp.

1945.   A Biological Spectrum of the Flora of the Great Smoky Mountains National Park. *Butler Univ. Bot. Studies* 7:11-24.

Cain, S. A. and L. R. Hesler
1940.   Harry Milliken Jennison, 1885-1940. *Jour. Tenn. Acad. Sci.* 15:173-176.

Cain, S. A. and A. J. Sharp
1938.   Bryophytic Unions of Certain Forest Types of the Great Smoky Mountains. *Amer. Midland Nat.* 20:249-301.

Cain, S. A. et al.
1937.   A Preliminary Guide to the Greenbrier-Brushy Mountain Nature Trail, the Great Smoky Mountains National Park. Botany Dept., Univ. Tenn. 43 pp. (mimeographed).

Camp, W. H.
1931.   The Grass Balds of the Great Smoky Mountains of Tennessee and North Carolina. *Ohio Jour. Sci.* 31:157-164.

1936.   On Appalachian Trails. *N. Y. Bot. Gard. Jour.* 37:249-265.

1938.   Studies in the Ericales III. The Genus *Leiophyllum. Bull. Torrey Bot. Club* 65:99-104.

1942a.   On the Structure of Populations in the Genus *Vaccinium*. *Brittonia* 4:189-204.

1942b.   A Survey of the American Species of *Vaccinium*, Subgenus *Euvaccinium*. *Brittonia* 4:205-247.

1945.   The North American Blueberries with Notes on Other Groups of Vacciniaceae. *Brittonia* 5:203-275.

1951.   The Biogeographic and Paragenetic Analysis of the American Beech *(Fagus)*. *Amer. Phil. Soc. Yrbk*. 1950:166-169.

Campbell, C. C.
1936.   Trees on Stilts. *Nature Notes* 1:56. Peoria, Ill.

Campbell, C. C. et al.
1962.   *Great Smoky Mountains Wildflowers*. Univ. of Tenn. Press, Knoxville. 40 pp. (illus. in color).

1964.   *Great Smoky Mountains Wildflowers Enlarged Edition*. Univ. of Tenn. Press, Knoxville. 88 pp. (illus. in color).

Coker, W. C. and H. R. Totten
1934.   *Trees of the Southeastern States*. Univ. of N. C. Press, Chapel Hill. 399 pp.

Copeland, H. F.
1943.   A Study, Anatomical and Taxonomic, of the Genera of Rhododendroideae. *Amer. Midland Nat*. 30:533-625.

Crandall, D. L.
1958.   Ground Vegetation Patterns of the Spruce-fir Area of the Great Smoky Mountains National Park. *Ecol. Monog*. 28:337-360.

Deam, C. C.
1940.   *Flora of Indiana*. Dept. of Conservation, Div. of Forestry, Indianapolis, Ind. 1236 pp.

Dixon, D.
1961.   These are the Champs. *Amer. Forests* 67, No. 1: 41-50; 67, No. 2:41-47.

Duncan, W. H. and T. M. Pullen
1962.   Lepidote Rhododendrons of the Southeastern United States. *Brittonia* 14:290-298.

Edwin, G.
1957.   Notes on *Ilex. Rhodora* 59:20-23.

Fernald, M. L.
1950.   *Gray's Manual of Botany;* a Handbook of the Flowering Plants and Ferns of the Central and Northeastern United States and Adjacent Canada. 8th ed. American Book Co., New York. 1632 pp.

Fink, P. M.
1931.   A Forest Enigma. *Amer. Forests*, 37:538, 556.

Fleetwood, R. J.
1934-35.   Journal of Raymond J. Fleetwood, Wildlife Technician, Great Smoky Mountains National Park, for the period May 27, 1934-June 27, 1935. 499 pp. (typewritten).

Flint, R. F.
 1947.   *Glacial Geology and the Pleistocene Epoch.* Wiley, New
                York. 589 pp.
Galle, F. C.
 1963.   Azaleas on Gregory Bald, Blount County, Tennessee. Re-
                port of 4 pp. dated Jan. 24 (typewritten).
Galyon, W. L.
 1928a.   The Smoky Mountains and the Plant Naturalist. *Jour.
                Tenn. Acad. Sci.* 3 (2):1-11.
 1928b.   Check List of the Trees and Shrubs of Eastern Tennessee.
                Thesis (M.A.) Univ. Tenn.
Ganier, A. F.
 1956.   Nesting of the Black-throated Blue and Chestnut-sided
                Warblers. *Migrant* 27:43-46.
Gates, W. H.
 1941.   Observations on the Possible Origin of the Balds of the
                Southern Appalachians. La. State Univ. Press, Baton
                Rouge *(Biol. Abstr.* 16:6295, 1942). 16 pp.
Gattinger, A.
 1887.   *The Tennessee Flora;* with Special Reference to the Flora
                of Nashville. Published by the author. Nashville, Tenn.
                109 pp.
 1901.   *The Flora of Tennessee.* Tenn. Bur. of Agric. Nashville,
                Tenn. 296 pp.
Gilbert, V. C., Jr.
 1954.   Vegetation of the Grassy Balds of the Great Smoky Moun-
                tains National Park. Thesis (M.S.) Univ. Tenn. 73 pp.
Gleason, H. A.
 1952.   *The New Britton and Brown Illustrated Flora of the North-
                eastern United States and Adjacent Canada.* N. Y. Bot.
                Garden. 3 vols.; 482, 655, and 589 pp.
Grimm, W. C.
 1957.   *The Book of Shrubs.* Stackpole Co., Harrisburg, Pa. 522
                pp.
Hardin, J. W.
 1954.   An Analysis of Variation within *Magnolia acuminata* L.
                Jour. Elisha Mitchell Sci. Soc. 70:298-312.
Harlow, W. M.
 1942.   *Trees of the Eastern United States and Canada.* Whittlesey
                House, New York. 288 pp.
Harper, R. M.
 1947.   Preliminary List of Southern Appalachian Endemics. *Cas-
                tanea (Jour. So. Appal. Bot. Club)* 12:100-112.
Hesler, L. R.
 1962.   List of Fungi of the Great Smoky Mountains National
                Park. (Revised Nov. 1, 1962.) Botany Dept., Univ. Tenn.
                83 pp. (typewritten).

Hoffman, H. L.
1962.   Check List of Vascular Plants of the Great Smoky Mountains National Park, Gatlinburg, Tenn. 44 pp. (mimeographed).

Holmes, J. S.
1911.   Forest Conditions in Western North Carolina. *N. C. Geol. & Econ. Surv. Bull.* 23:1-111.

Hu, S.
1954-56.   A Monograph of the Genus *Philadelphus. Jour. Arnold Arboretum* 35:275-333 (1954); 36:52-109, 325-368 (1955); 37:15-90 (1956).

Hunt, G. E.
1962.   Royal E. Shanks (1912-1962). *Jour. Tenn. Acad. Sci.* 37:110.

Illick, J. S.
1928.   Pennsylvania Trees. *Bull. No. 11,* Pa. Dept. Forests and Waters. 237 pp.

Jennison, H. M.
1935a.   Notes on Some Plants of Tennessee. *Rhodora* 37:309-323.
1935b.   Trees of the Great Smoky Mountains Park. Special report accompanying his November 1935 monthly report to the Superintendent, Great Smoky Mountains National Park. 12 pp. (typewritten).
1938.   A Classified List of the Trees of Great Smoky Mountains National Park. Typewritten report to Great Smoky Mountains National Park. 26 pp.
1939a.   A Preliminary Catalog of the Flowering Plants and Ferns of the Great Smoky Mountains National Park. Typewritten report to Great Smoky Mountains National Park. 249 pp.
1939b.   Flora of the Great Smokies. *Jour. Tenn. Acad. Sci.* 14:266-298.

Jones, G. N.
1946.   American Species of *Amelanchier. Ill. Biol. Mon.* 20:2...

Journal of Heredity
1915.   Photographs of large trees. Vol. 6, No. 9, Sept. 1915; pp. 407-423 (no author).

Kelsey, H. P.
1949.   Unique Flora of the Great Smoky Mountains National Park. *Arboretum Bull.* 12:11-13, 30.

Kelsey, H. P. and W. A. Dayton
1942.   *Standardized Plant Names.* 2nd ed. J. Horace McFarland Co., Harrisburg, Pa. 675 pp.

King, P. B. and Arthur Stupka
1950.   The Great Smoky Mountains—Their Geology and Natural History. *Sci. Monthly* 71:31-43.

Korstian, C. F.
1937.   Perpetuation of Spruce on Cut-over and Burned Lands in

the Higher Southern Appalachian Mountains. *Ecol. Monog.* 7:125-267.

Lambert, R. S.

1958, 1960.   Logging in the Great Smoky Mountains National Park. A report to the Superintendent, Great Smoky Mountains National Park. 71 pp. (typewritten) plus maps and miscellaneous items.

1961.   Logging on Little River, 1890-1940. *E. Tenn. Hist. Soc. Publ. 33:1-12* (Knoxville, Tenn.).

Little, E. L., Jr.

1953.   Check List of Native and Naturalized Trees of the United States (Including Alaska). *U. S. Dept. Agr. Handbook* 41: 1-472.

Maddox, R. S.

1926.   The Trees of the Great Smokies. *Jour. Tenn. Acad. Sci.* 1:21-24.

Mark, A. F.

1958.   The Ecology of the Southern Appalachian Grass Balds. *Ecol. Monog.* 28:293-336.

McAtee, W. L.

1956.   *A Review of the Nearctic Viburnum.* Publ. by the author. Chapel Hill, N. C. 125 pp.

Miller, F. H.

1938.   Brief Narrative Descriptions of the Vegetative Types in the Great Smoky Mountains National Park. Report to the Superintendent, Great Smoky Mountains National Park. 17 pp. (typewritten).

1941.   Vegetation Type Map, Great Smoky Mountains National Park. In color: 27 x 60 inches. Civilian Conservation Corps Project (1935-1938).

Munns, E. N.

1938.   The Distribution of Important Forest Trees of the United States, *USDA Miscell. Pub. 287.* Washington, D. C. 176 pp.

Oakes, H. N.

1932.   *A Brief Sketch of the Life and Works of Augustin Gattinger.* Cullom & Ghertner Co., Nashville, Tenn. 152 pp.

Oosting, H. J. and W. D. Billings

1951.   A Comparison of Virgin Spruce-fir Forest in the Northern and Southern Appalachian System. *Ecology* 32:84-103.

Ramseur, G. S.

1959.   A Natural Stand of Rhododendron X Wellesleyanum Waterer ex Rehder in the Southern Appalachians. *Jour. Elisha Mitchell Sci. Soc.* 75:131.

1960.   The Vascular Flora of High Mountain Communities of the Southern Appalachians. *Jour. Elisha Mitchell Sci. Soc.* 76: 82-112.

Russell, N. H.
    1953.   The Beech Gaps of the Great Smoky Mountains. *Ecology*
              34:366-374.
Sargent, C. S.
    1933.   *Manual of the Trees of North America.* Houghton Mif-
              flin Co., Boston. 910 pp.
Schofield, W. B.
    1960.   The Ecotone between Spruce-fir and Deciduous Forests in
              the Great Smoky Mountains. Thesis (Ph.D.), Duke Univ.,
              Durham, N. C.
Shanks, R. E.
    1947.   Key to the Species of *Vaccinium* in the Great Smoky
              Mountains (adapted from Camp, 1945). 1 p. (typewritten).
    1952.   Checklist of the Woody Plants of Tennessee. *Jour. Tenn.
              Acad. Sci.* 27:27-50.
    1953.   Great Smoky Mountains National Park *Crataegus* (based
              on determinations made by E. J. Palmer in August and
              Sept. 1953), 1 p. (penciled).
    1954a.   Reference Lists of Native Plants of the Great Smoky
              Mountains. Botany Dept., Univ. Tenn. 17 pp. (mimeo-
              graphed).
    1954b.   Plotless Sampling Trials in Appalachian Forest Types.
              Ecology 35:237-244.
    1954c.   Climates of the Great Smoky Mountains. *Ecology* 35:354-
              361.
    1956.   Altitudinal and Microclimatic Relationships of Soil Tem-
              perature under Natural Vegetation. *Ecology* 37:1-7.
    1957.   Provisional Key to Tennessee Species of *Rubus*. Botany
              Dept., Univ. Tenn. 6 pp. (mimeographed).
    1961.   Woody Plants of the Great Smoky Mountains. Botany
              Dept., Univ. Tenn. 19 pp. (typewritten).
Shanks, R. E. and A. J. Sharp
    1947.   Summer Key to the Trees of Eastern Tennessee. *Jour.
              Tenn. Acad. Sci.* 22:114-133.
    1950.   *Summer Key to Tennessee Trees.* Botany Dept., Univ.
              Tenn. Reprinted in 1963 by Univ. Tenn. Press, Knoxville.
              24 pp.
Sharp, A. J.
    1942a.   A Preliminary Checklist of the Trees in the Great Smoky
              Mountains National Park. (Revised by F. H. Arnold in
              1945.) Typewritten report to Great Smoky Mountains Na-
              tional Park. 39 pp.
    1942b.   A Preliminary List of the Woody and Semi-Woody Shrubs
              and Vines Occurring in Great Smoky Mountains National
              Park. (Revised by F. H. Arnold in 1945.) Typewritten
              report to Great Smoky Mountains National Park. 17 pp.
    1957.   Vascular Epiphytes in the Great Smoky Mountains. *Ecol-
              ogy* 38:654-655.

Sharp, A. J. et al.
1956.   A Preliminary Checklist of Monocots in Tennessee. Botany
            Dept., Univ. Tenn. 33 pp. (mimeographed).
1960.   A Preliminary Checklist of Dicots in Tennessee. Botany
            Dept., Univ. Tenn. 114 pp. (mimeographed).
Shaver, J. M.
1926.   Flowers of the Great Smokies. *Jour. Tenn. Acad. Sci.*
            1:17-20.
Skinner, H. T.
1955.   In Search of Native Azaleas. *Morris Arboretum Bull.* 6:3-
            10, 15-22.
Small, J. K.
1933.   *Manual of the Southeastern Flora.* Small, New York.
            1554 pp.
Society of American Foresters
1932.   *Forest Cover Types of the Eastern United States.* Wash-
            ington, D. C. 48 pp.
Stupka, Arthur
1935-62.   Nature Journal, Great Smoky Mountains National Park.
            28 vols. (years) each with index (typewritten).
1960.   *Great Smoky Mountains National Park, Natural History
            Handbook, Series No. 5.* U. S. Government Printing Of-
            fice, Washington, D. C. 75 pp.
1963.   *Notes on the Birds of Great Smoky Mountains National
            Park.* Univ. of Tenn. Press, Knoxville. 256 pp.
Totten, H. R.
1937.   Notes on *Buckleya* and *Pyrularia* (Buffalo-nut). *Jour. Elisha
            Mitchell Sci. Soc.* 53:226.
U. S. Dept. of the Interior
1959.   Great Smoky Mountains National Park, North Carolina
            and Tennessee (information booklet). U. S. Government
            Printing Office, Washington, D. C. 32 pp.
Wells, B. W.
1936a.   Origin of the Southern Appalachian Grass Balds. *Science,*
            N.S. 83:283.
1936b.   Andrews Bald: The Problem of Its Origin. *Castanea
            (Jour. So. Appal. Bot. Club)* 1:59-62.
1937.   Southern Appalachian Grass Balds. *Jour. Elisha Mitchell
            Sci. Soc.* 53:1-26.
1946.   Archeological Disclimaxes. *Jour. Elisha Mitchell Sci. Soc.*
            62:51-53.
1956.   Origin of Southern Appalachian Grass Balds. *Ecology* 37:
            592.
Whittaker, R. H.
1952.   A Study of Summer Foliage Insect Communities in the
            Great Smoky Mountains. *Ecol. Monog.* 22:1-44.
1956.   Vegetation of the Great Smoky Mountains. *Ecol. Monog.*
            26:1-80.

Woods, F. W. and R. E. Shanks
    1957.   Replacement of Chestnut in the Great Smoky Mountains of Tennessee and North Carolina. *Jour. Forestry* 55 (11): 847.
    1959.   Natural Replacement of Chestnut by other Species in the Great Smoky Mountains National Park. *Ecology* 40 (3): 349-361.

Species Index ———————————

This volume is published
with the cooperation of the
Great Smoky Mountains Natural History Association